# A Trillion Trees

FRED PEARCE

# A Trillion TREES

## Restoring Our Forests by Trusting in Nature

GREYSTONE BOOKS
Vancouver/Berkeley/London

North American edition first published by Greystone Books in 2022
Originally published in Great Britain by Granta Books in 2021

22 23 24 25 26 5 4 3 2 1



Greystone Books Ltd.
greystonebooks.com

Cataloguing data available from Library and Archives Canada
ISBN 978-1-77164-940-7 (cloth)
ISBN 978-1-77164-941-4 (epub)

Copy editing for North American edition by Paula Ayer
Proofreading by Jennifer Stewart
Jacket and text design by Belle Wuthrich
Jacket illustration by Shutterstock/Nosyrevy
Background photograph by Adrian Pingstone
Maps © John Gilkes
Printed and bound in Canada on FSC® certified paper at Friesens. The FSC® label means that materials used for the product have been responsibly sourced.

Greystone Books gratefully acknowledges the Musqueam, Squamish, and Tsleil-Waututh peoples on whose land our Vancouver head office is located.

Greystone Books thanks the Canada Council for the Arts, the British Columbia Arts Council, the Province of British Columbia through the Book Publishing Tax Credit, and the Government of Canada for supporting our publishing activities.

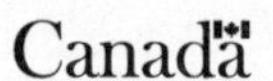

*To the memories of Don Hinrichsen*
*and Conrad Gorinsky, pioneers both*

# CONTENTS

x *Map: A Forest Journey*
1 **Introduction: Myth and Magic**

PART I
WEATHER MAKERS

17 **1. Trees Are Cool**
Stomata, Transpiration and a Planet Transformed

30 **2. Flying Rivers**
Chasing the Rain and Mapping a New Hydrology

43 **3. Forests' Breath**
Sniffing the Air and Shooting the Breeze

59 **4. In Tanguro**
Tipping Points in Soybean Fields Foreshadow Crisis in the Amazon

72 **5. Fires in the Forest**
Nature's Way of Starting Over

## PART II
## FROM PARADISE TO PLUNDER

85 **6. Lost Worlds**
Pre-Columbian Cities That Gardened the Rainforests

97 **7. The Woodchopper's Ball**
Post-Columbian Pillage and Roads to Ruin

109 **8. Logged Out**
Well, Almost... Three Decades in Borneo

121 **9. Consuming the Forests**
Logs of War and a New "Green" Plunder

135 **10. No-Man's-Land**
Cattle Kingdoms and the Tyranny of Global Commodities

144 **11. Taking Stock**
Phantom Forests and Debunking Forest Demonology

## PART III
## REWILDING

161 **12. From "Stumps and Ashes"**
America's Forest Renaissance

168 **13. The Strange Regreening of Europe**
Acid Rain to a New Green Deal

179 **14. Forest Transition**
How More and More Nations Are Restoring Their Forests

187 **15. To Plant or Not to Plant**
When Trees Become Part of the Problem

201 **16. Let Them Grow**
Only Nature Can Plant a Trillion Trees

210 **17. Agroforests**
Farmers as Part of the Solution

## PART IV
## FOREST COMMONS


227 **18. Indigenous Defenders**
Why Tribes Do Conservation Better Than Conservationists

244 **19. Community Forests**
A Triumph of the Commons

257 **20. African Landscapes**
Taking Back Control

271 **Postscript: Back Home**

279 Acknowledgments
283 Notes
315 Further Reading
321 Index

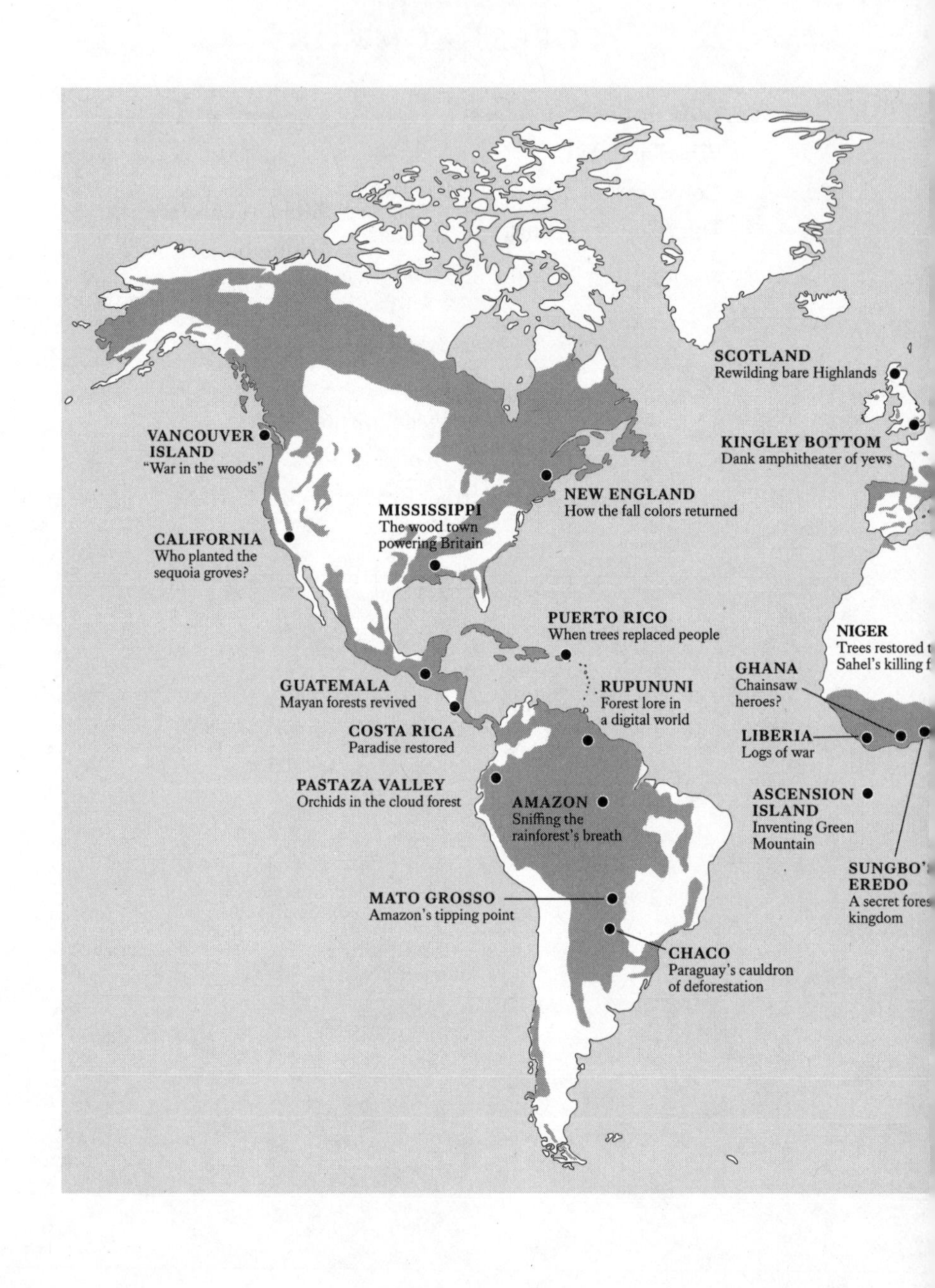

SCOTLAND
Rewilding bare Highlands
VANCOUVER ISLAND
"War in the woods"
KINGLEY BOTTOM
Dank amphitheater of yews
NEW ENGLAND
How the fall colors returned
MISSISSIPPI
The wood town powering Britain
CALIFORNIA
Who planted the sequoia groves?
PUERTO RICO
When trees replaced people
NIGER
Trees restored t
Sahel's killing f
GHANA
Chainsaw heroes?
GUATEMALA
Mayan forests revived
RUPUNUNI
Forest lore in a digital world
COSTA RICA
Paradise restored
LIBERIA
Logs of war
PASTAZA VALLEY
Orchids in the cloud forest
AMAZON
Sniffing the rainforest's breath
ASCENSION ISLAND
Inventing Green Mountain
SUNGBO'
EREDO
A secret fores
kingdom
MATO GROSSO
Amazon's tipping point
CHACO
Paraguay's cauldron of deforestation

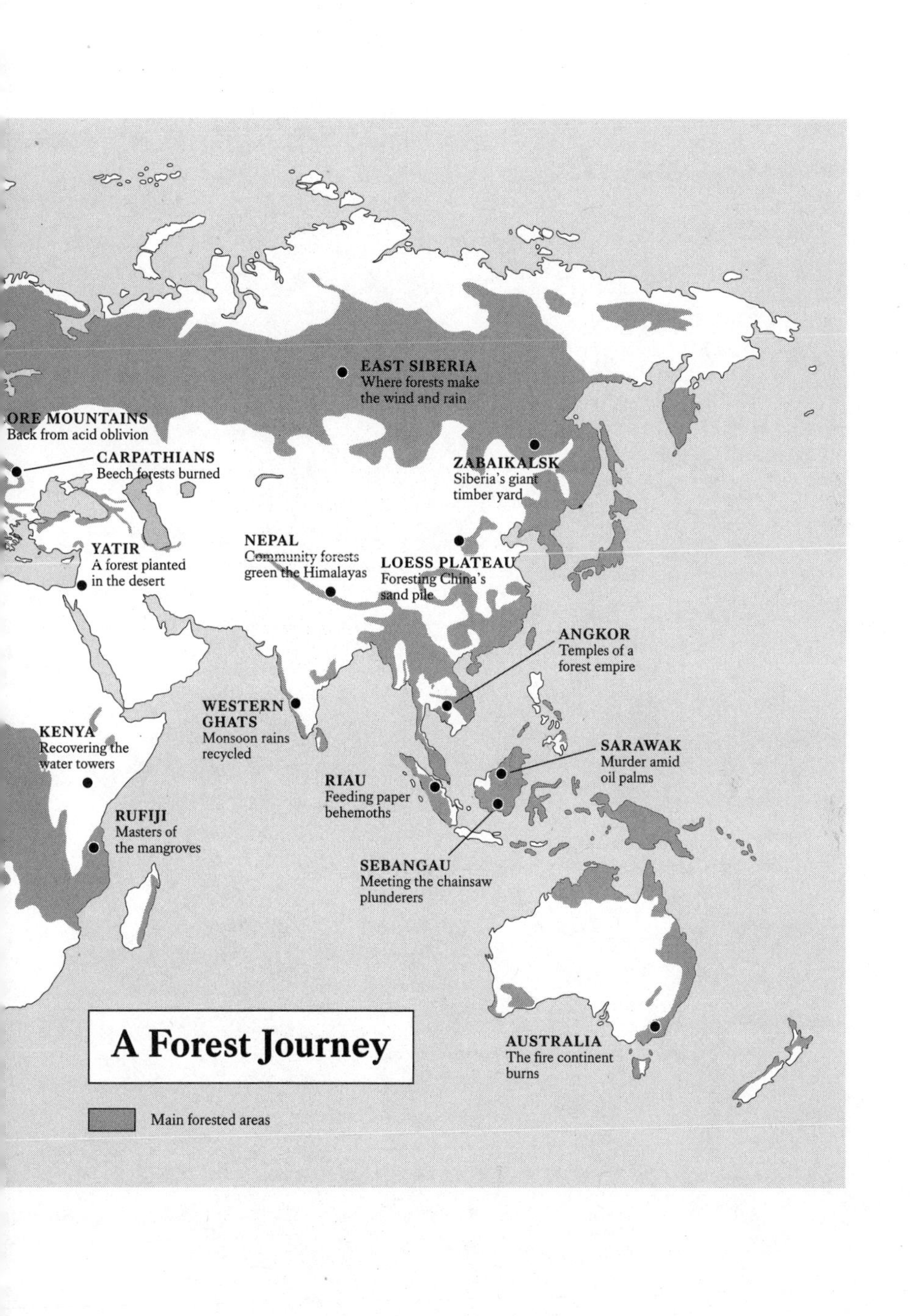

EAST SIBERIA
Where forests make the wind and rain
ORE MOUNTAINS
Back from acid oblivion
CARPATHIANS
Beech forests burned
ZABAIKALSK
Siberia's giant timber yard
YATIR
A forest planted in the desert
NEPAL
Community forests green the Himalayas
LOESS PLATEAU
Foresting China's sand pile
ANGKOR
Temples of a forest empire
WESTERN GHATS
Monsoon rains recycled
KENYA
Recovering the water towers
SARAWAK
Murder amid oil palms
RIAU
Feeding paper behemoths
RUFIJI
Masters of the mangroves
SEBANGAU
Meeting the chainsaw plunderers
A Forest Journey
AUSTRALIA
The fire continent burns
Main forested areas

# INTRODUCTION

## *Myth and Magic*

MY FIRST REAL EXPERIENCE of the wonder of tropical forests came in the Andean Mountains of Ecuador. It had been a long drive from Quito and evening was drawing on as we climbed into forests shrouded in clouds—clouds whose coverage was so extensive and permanent that no cartographers had ever mapped the terrain, and no satellites had ever observed the surface beneath. Breathing the sopping-wet air and peering into the gloom, I could understand why people occasionally showed up here believing that the trees hid an El Dorado of gold buried by the Inca half a millennium ago.

But I hadn't come for gold. Walking the next morning into the eternal mists above the valley town of Baños, I met Lou Jost. An American botanical explorer, he had come here to uncover his own El Dorado: a biological trove of orchids found nowhere else. He said his discoveries were changing our understanding of how and why plants evolve to create species that are unique to particular places, and why forests contain so much of the world's biodiversity.

When I met him, Jost had already spent six years exploring the ridges around Baños. He operated alone, without the help of any academic body, having given up work as a quantum physicist in the

US to pursue his dream. He said he found scouring the cloud forests for orchids more mind-expanding than the wonders of the universe. "Quantum physics offered nothing you could touch and feel," he told me. "But as soon as I visited a cloud forest, I was hooked." On the wet, sunless forest floor, species of orchids have evolved in the thin air with flowers so fragile that they would collapse anywhere else. He had already discovered ninety previously unknown orchids, and believed there were more here than anywhere else in the world. He had become an artist too, drawing and painting his discoveries.[1]

The Pastaza valley, where Baños is situated, is the deepest and straightest valley in the eastern Andes, draining down into the vast basin of the Amazon River to the east. Despite being virtually on the equator, it is bone-chillingly cold and wet, and riddled with cliffs and ravines that are all the more dangerous in the permanent cloud. Not many people pass this way. The infrequent trails are mostly made by the mountain tapirs and spectacled bears.

The outside world intervenes, nonetheless. Every day a wind blows in from the Amazon rainforest, bringing huge volumes of moisture that condenses to form the near-permanent clouds. "Each ridge catches the winds a little differently and has a different microclimate," said Jost. "Each species seems to specialize in a particular combination of rain, mist, wind and temperature." Some species grow by the thousand on top of a single ridge, but disappear just a few feet below and are found nowhere else.

"The only way to discover the botanical secrets of these forests is to walk them all," Jost told me. Many of the hills have never been visited by scientists. Well, almost never. Jost came originally to follow in the footsteps of his hero, English botanist Richard Spruce, who trekked through the Pastaza valley in the 1850s, discovering ferns and liverworts that have never been seen since. The valley is comparable to Ecuador's other, more famous biological treasure house, the Galápagos Islands. On one red-letter day, Jost found four new species in a single patch of moss, raising the number of known orchid species of the genus *Teagueia* from six to ten. Since then, he has identified more of the distinctive long-creeping orchids here. One, *Teagueia jostii*, bears his name.

Jost's orchid collection had become a life-consuming passion. His apartment in Baños was strewn with plant samples. A tiny rooftop greenhouse grew plants collected on past expeditions. Many had only opened their petals under his tender care. "You have to know what you are looking for when you go orchid hunting," he says. "The flowers are only a few millimeters across and usually hide under the leaves. Often the plants are not in flower. If I spot what I think is a new species, I can often only be sure when I bring it back here to wait for the flower to appear."

Their survival in Jost's apartment looked precarious. He kept the greenhouse air mountain-cool with an electric fan, which depended on the town's fitful power supply. As a backup, he had a passive air conditioner that drew in air over permanently wet tiles. There was an ingenious device to maintain humidity, using a paperweight balance. When a packet of plant stalks on the balance became too dry, it lost weight and the balance shifted, switching on a humidifier. Once moist again, it switched the humidifier off. "It's not perfect. But it means I can go away on plant-collecting expeditions and be fairly sure my specimens will still be alive when I return," he said.

Can Jost's wild orchid El Dorado survive? The value of the Baños cloud forests' biological heritage is increasingly being acknowledged. In 2006, near a waterfall just down the valley from Baños, Jost and local officials erected a bust to commemorate Spruce's journey there. Jost, a loner when I visited him, is becoming a celebrity himself. He now runs the EcoMinga Foundation, an NGO dedicated to protecting Ecuador's upland ecosystems. In 2016, TV naturalist David Attenborough launched a film about Jost and "the orchids of Baños."

Tourists are flocking to Baños too. Some come to kayak or go white-water rafting on the Pastaza River as it cascades down towards the Amazon. Others are scientists and nature buffs. Their enthusiasm is changing local perceptions about the cloud forests. "In the past, the *campesinos* saw nature and the environment as a symbol of their poverty. But scientists coming here have introduced them to the idea that it is something to be proud of," said Patricia Guevara, a former vice-mayor of Baños who spotted the trend early and set up an eco-lodge in the valley. Jost too sees the growing interest as good news. "These kinds of tourists bring money and fill hotels, but to keep them, you have to keep nature as well." He should know. He was a tourist once too, lured by the legend of Spruce and, staying on, captivated by the magic of the orchids of the cloud forests.

THE WORLD'S GREAT FORESTS are awe-inspiring places, but some of my most memorable forest experiences have happened on my own doorstep. I remember, at the age of about seven, getting lost with a friend in a wood near my home on the North Downs in Kent. It was a small wood, but we kept walking in circles. After an hour, the sun was setting. There seemed to be no way out. My friend said a plane had crashed here during the Battle of Britain. There was a big hole that looked like the impact crater. We wondered if the ghost of the pilot might still be there. We were thoroughly spooked. It was probably the first time that I had contemplated my own mortality.

Forests have always transfixed the human imagination. Our fairy tales are full of cautionary tales about these dank, mysterious places. Little Red Riding Hood was waylaid in the woods by a big bad wolf as she delivered provisions to her grandmother. Hansel and Gretel lost their way and were kidnapped far from home by a cannibalistic witch. The fair maid Rapunzel was incarcerated in a tower deep in a wood. Even modern tales often invoke forests for their imagery of mystery, evil and dread. Think *The Lord of the Rings*, *Where the Wild Things Are* and *The Blair Witch Project*. Woods are where bodies are buried, darkness pervades and evil lurks. Swiss psychoanalyst Carl Jung called their hold on us "primordial."

But there is wonder too. I also remember childhood walks in the cool air and dappled light of Wealden woods. As an adult, I return repeatedly to the ancient yews of Kingley Bottom in southern England, Europe's largest and oldest yew forest. Surrounded by grassy downland and overlooked by Bronze Age burial mounds, it speaks to me of the permanence of humans in natural landscapes.

Most of the world's religions use trees as symbols of life. The Koran features a tree of immortality, where the souls of righteous people live forever. Christians and Jews share a tree of knowledge, the centerpiece of the story of the fall of mankind in the Garden of Eden. The branches on the Buddhists' tree of wisdom, a fig growing in the Mahabodhi Temple in Bihar, India, are the source of the four rivers of life. Such reverence is not surprising. As British naturalist Richard Mabey points out in his book *The Cabaret of Plants*, trees are "capable of outliving not just individual humans but whole civilizations."[2]

Trees have helped shape civilizations too, inspiring a colonial lust to explore and tame. When the early European explorers encountered the rainforests of the tropics, their imaginations were infected with both fear and excitement. From Francisco de Orellana, the first European to travel the length of the Amazon, to Sir Walter Raleigh, the courtier to Queen Elizabeth who twice sailed up the Orinoco, the Spaniards and English for centuries went to South America in search of El Dorado, a city hidden in the jungle that they believed to be full of gold.

There was awe. Raleigh called the forests he sailed through "a country that hath yet her maidenhead, never sacked, turned nor wrought."[3] Two centuries on, Prussian scientist-explorer Alexander von Humboldt found "an inexhaustible treasure trove... just like paradise." The rainforests were places where, he wrote, "every object declares the grandeur of the power, the tenderness of nature, from the boa constrictor which can swallow a horse, down to the hummingbird balancing itself on the chalice of a flower." He saw their inhabitants in similar terms: an "infant society" that "enjoyed pure and perpetual felicity."[4]

Those explorers also found a world so alien to them that they returned home telling tales of inexplicable horror. Many recounted stories of an Amazonian tribe whose members had no heads, but eyes in their shoulders and mouths on their breasts, and of warrior women who procured and discarded slave men at will. In Africa, the British explorer Henry Morton Stanley called the rainforest "a murderous world" full of "filthy vulturous ghouls."[5]

Some nineteenth-century explorers found time to observe more dispassionately. Jost's hero Richard Spruce, a shy, sickly math teacher from Yorkshire in England, spent fifteen years collecting ferns, mosses, lichens and liverworts in the Amazon basin. In his notebooks, he analyzed a "tendency to gigantism" in the jungle, with bird-eating spiders, eight-inch slugs, seven-inch butterflies and goliath beetles. But even he succumbed to the almost psychedelic intensity of this world, summed up in his sampling of the botanical hallucinogen *Ayahuasca*, known as the "vine of the soul." First, it gave him visions of "beautiful lakes, woods laden with fruit, birds of brilliant plumage." Then paradise became purgatory, as his body "turned deadly pale, trembling in every limb, bursting with perspiration and seeming possessed with reckless fury."[6]

Even today, for those who don't dwell in them, forests remain somehow "other." German filmmaker Werner Herzog, in *Burden of Dreams*, saw rainforests as places of "overwhelming misery and overwhelming fornication, overwhelming growth and overwhelming lack of order."[7] Forests are places that still seem to exist apart from global events, their occupants sometimes losing touch with reality. There are stories of Second World War combatants emerging from a jungle decades later, unaware that hostilities were over. A Russian family that fled Stalin's purges in the 1930s was found forty years later, holed up in Siberian forests, unaware of Stalin's death or even of the Second World War.[8]

These days we more often accentuate the positive. The new El Dorado is primarily biological. Fearful "jungles" are rebranded as bountiful rainforests, miracles of biodiversity where nature makes chemicals that can cure cancer. A sense of sheer wonder persists, even among our most eminent forest scientists. The biodiversity guru Edward Wilson calls rainforests "timeless, immutable... the crucible of evolution."[9] For climatologists they are the "lungs of the planet."

Despite scientists' attempts to discover the forests' secrets, many mysteries remain. This book seeks to investigate some of them and help us see forests a little more clearly. One unheralded truth is that most of the world's forests—even the largest and most remote—are not as old or as wild as we like to imagine. There may be some exceptions: perhaps a ridge or two in the cloud forests of Ecuador, or some remote portion of the vast northern forests of Siberia, or maybe even a fraction of the boggy peat-swamp forests of Southeast Asia could have escaped our attentions. But almost all forests are marked by extensive human occupation and alteration that are now imprinted in their ecology. No forest still "hath her maidenhead."

Many of the wildest, apparently least-tamed forests are little more than overgrown gardens created by the ancestors of those who live in them today. That, I believe, has important lessons for how we should treat our forests and their inhabitants today. We may indulge our forest fantasies. Even so, we should remember they are at least as much human landscapes as pristine paradise.

For decades, I have been reporting on the importance of trees—for keeping us cool, preserving the planet's species, maintaining rainfall, alleviating poverty and protecting Indigenous people. I have written too about who owns them, who uses them, who protects them and who trashes them. This book is an extension of that work. It is about the magic and mystery of trees and forests, about their defenders and plunderers, and why they matter for the planet and for all of us. Along the way, it will chart the extraordinary pace of the forests' destruction, which peaked at the end of the twentieth century, but also explore where and why they are recovering. It will look forward to a great forest restoration in the coming decades, explain how it can happen and why it must.

It will also take you on a journey through the world's forests—to the forty or more countries I have visited to enter forests and interview their inhabitants. To the cloud forests of the Ecuadorian Andes, which few outsiders have explored since Spruce, and the radioactive (but otherwise healthy) forests around Chernobyl in Ukraine; to the swamp forests of Indonesia and the community forests of the Himalayas; to the acid-rain-ravaged forests of central Europe and the pine forests in the American Deep South being cut to keep the lights on in Britain; to sacred groves in India and forests that funded war in Liberia; to the depths of the Amazon and the peak of Ascension Island's Green Mountain in the middle of the Atlantic Ocean; to the vast larch and pine forests of Siberia and the multicolored magnificence of New England in the autumn.

In my travels, I have flown for hours over the hundreds of billions of trees in the Amazon, and got up close to the only tree on the windswept Falkland Islands—in the British governor's garden. I have visited the remains of a largely unknown ancient forest civilization in Nigeria and the remarkable natural reforesting of the drought lands of the African Sahel. I have penetrated the weird thorn forests of the Paraguayan Chaco, under siege from Brazilian ranchers, and seen China's attempts to turn back its deserts with trees. I have visited the front line of the palm oil invasion in Borneo and the desecration of Sumatra's forests to keep paper in your printer. I have toured a forest on the edge of the Israeli desert that is perversely adding to global warming and climbed a gantry as high as the Eiffel Tower in the middle of the Amazon rainforest, erected so German scientists can sniff the breath of the planet's "lungs."

I have met Scottish crofters reforesting the Highlands and Indigenous tribes mapping their forests with GPS on their phones; a Russian nuclear physicist who believes that forests make the world's winds and an English adventurer trying to prove her right in his back garden; a playboy bush pilot who showed the world that the Amazon forests water a "flying river" that makes rain; an Israeli who says planting a billion trees in the Sahara could make it as lush as the Amazon; a hired farmhand in Malaysia facing execution for assassinating a forest campaigner; and an American scientist in exile after being pilloried for suggesting that forests could sometimes be bad for the climate. Back home I reconnected with the gnarled yews of Kingley Bottom and the bluebell woods and colossal sweet chestnuts of nearby Burton Park. Each of these encounters with forests, and with those who know them best, has helped me understand the significance of trees for the continuation of our planet and of human life.

ABOUT HALF THE WORLD'S forests have gone since the dawn of civilization. What we have lost shames us as a species. The sound of chainsaws has replaced birdsong from the steppes of Russia to the backwoods of Australia. But what we have left is immensely precious. Almost a third of the planet's land surface is still covered by around three trillion trees—more than all the stars in the Milky Way.[10] They are home to more than half the world's species. They cleanse air and water. They deliver fruits and nuts, rubber and timber, honey and medicines. They manage the water cycle, storing water in soil to maintain river flows and control floods. They control the climate too. They store as much carbon as humans have emitted into the atmosphere since the start of the Industrial Revolution. And each day, every tree pumps around thirteen gallons of water into the atmosphere from quintillions of tiny pores on their leaves, keeping the atmosphere moist and cool.

The good news is that the great shaving of our planet may be coming to a close. In some countries, humans are putting back trees. Sometimes by planting. China and India, the world's two most populous nations, have found room to seed more forests. From Costa Rica to Nepal, there are similar stories. However, often—and just as importantly—we are giving natural forests room to recover. For the other great unheralded truth about our forests is that they are far from being passive victims of our assaults. Everywhere they fight back to restore their domain.

For several years, there has been a growing campaign to restore an additional trillion trees by, say, the end of this century, to help fight climate change, restore ecosystems and halt species extinctions. The idea is both beguiling and doable. Recent research says there is room—much of it on formerly forested lands that have not been turned into farms or cities or mines or anything else. Some say we could plant up parts of our great grasslands too. Deserts even.

I buy the ambition. A planet with a trillion more trees would be a much better place. My problem is not with the ambition of a trillion more trees; it is with the word "planting." It implies, indeed requires, a global industry, taking over farms and former forest land, often riding roughshod over local rights, in the name of reforestation. And it would be costly: almost $400 billion a year for the next thirty years, according to one assessment—a cool $12 trillion.[11] If we want a trillion more trees on our planet, as I believe we should, the last thing we need is a big planet-wide project to go out and plant them. It would be bad for people, bad for forests and in the end bad for the planet too. It would also be entirely unnecessary. We don't have to do the planting. We shouldn't do the planting. Nature will mostly do it for us. And she will do it better. If we stand back and give them room, forests will regrow.

Europe has a third more trees today than it had in 1900. So does North America. Most of them were not planted. Our planet as a whole has more trees than it did a decade ago. The great forest restoration may already be underway. Take a look at Niger, one of the world's most arid, poorest and hungriest nations. Farmers there have encouraged buried roots to repopulate their desiccated fields and been rewarded with better grain yields, richer soils and healthier families. A region once thought to be succumbing to the advancing Sahara is now greened, wetted and cooled by 200 million extra trees. If they can restore natural treescapes there, it can be done almost anywhere.

There will be no return to primeval wilderness. The world has changed too much for that, and in truth it is many centuries since there have been significant numbers of pristine forests. That may sound dispiriting, but actually it should give us hope. It should not extinguish the magic, for it shows that nature can everywhere be resilient and resurgent. The great forests have recovered from human activity before. They can again.

From the journeys I have made through the world's forested, deforested—and reforested—landscapes, the lesson seems to be consistent: we can put our trowels and bulldozers away. Pack up the seed nurseries and put our money back in our pockets. In most places, to restore the world's forests we need to do just two things: ensure that ownership of the world's forests is vested in the people who live in them, and give nature room. Rewilding the Earth does not mean replanting; it means the exact opposite.

THIS BOOK IS DIVIDED into four parts. In the first, "Weather Makers," I explore some new and extraordinary science about how fundamental forests are to our planet's life-support systems, and how trees have literally made the environment in which they—and we—prosper. This is not just to do with the carbon they store to curb the greenhouse effect of global warming, important though that is. It is also to do with how forests make the rain that sustains them, and how the forests' chemical "breath" helps make clouds and may even make the winds. It reveals how these systems are self-sustaining. The forests make the rain and the rain maintains the forests—though only up to a point. Beyond that tipping point, forests may rapidly degrade and disappear. We visit the front line where that tipping point may be being breached right now in the Amazon rainforests.

Part Two, "From Paradise to Plunder," asks how we got to this tipping point. Once, humanity had a good working relationship with forests. We mostly harvested them without destroying them. Yet in more recent times, we have forgotten how to do that, and instead felled half our forests and pushed the world towards a climatic Armageddon. During thirty years of environmental reporting I have seen the carnage created by soybean farming, palm oil plantations and cattle ranchers. But there is good news even here. The history of human occupation of our forests—including those we today regard as pristine—shows that deforestation need not be forever. Provided that they are not pushed too far, forests can and do recover from our depredations. Much of the Amazon is actually regrowth from clearance by former civilizations. So we need to take stock rather than be defeatist. Give them the chance and many forests can return.

In Part Three, "Rewilding," we discover that forest recovery is already happening. Europe and North America are much more forested than they were a hundred and fifty years ago, or a hundred years ago, or even fifty years ago. Sometimes those forests are not much like the old ones—especially where they have been planted. In most places, however, nature is already reclaiming her own, spreading new forests across abandoned fields. From the downs of southern England to the Russian steppes, from New England to the Deep South, we explore this brave new world of natural forest restoration and the surprising return of trees to farms. This rewilding is surely the new environmental agenda for the twenty-first century.

And, as I uncover in the final part, "Forest Commons," this return of the trees goes with the grain of an emerging community-centered approach to nature, forests and the land. We are relearning what we should never have forgotten: that forest people—whether Indigenous Amazonians or Nepalese hill dwellers, Kenyan farmers or Mexican peasants, Native Americans or West African sawyers—are the best custodians and conservators of existing forests and the best at giving new forests room to grow.

Let me say this clearly. I have been writing about the world's environmental problems for forty years—about toxic dumps and the ozone layer, deforestation and our emptying oceans, urban smogs and species extinction, climate change and spreading deserts. Despite all this, I remain an optimist. For the world can turn. The forests can regrow. Join me on my journey through the world's forests, past, present and—I profoundly believe—future.

PART I

# WEATHER MAKERS

---

Before the existence of forests, the atmosphere on Earth was baking hot, bone dry, short of oxygen and thick with carbon dioxide. Today, three trillion trees keep us cool and watered, by soaking up the carbon dioxide and by sweating moisture to sustain "flying rivers" that deliver rain across the world. Their breath alters atmospheric chemistry too, making clouds and even generating the winds. Trees, in short, created and sustain the life-supporting climate of our planet. Here is their story.

---

# —1—

# TREES ARE COOL

## *Stomata, Transpiration and a Planet Transformed*

TREES ARE THE BIGGEST and longest-living organisms on the planet. They can grow up to three hundred feet tall, weigh more than a thousand tons, and, in the case of North American bristlecone pines, sometimes live for more than four thousand years. In fact, their ability to clone replicas of themselves means they can effectively live forever. One cluster of forty thousand cloned aspens—or should that be aspen—in the Fishlake National Forest in Utah is a single male organism covering around a hundred acres and weighing more than six thousand tons. It is at least eighty thousand years old, making it "probably the oldest mass of connected tree tissue" on the planet, says British naturalist Richard Mabey. It hasn't grown in a while, so it may be dying.[1] Nobody can be sure of that either.

Still, trees are everywhere, and have been for hundreds of millions of years. Cloud forests cling to mountaintops, mangroves dangle their roots in tropical coastal waters, steaming jungles straddle the tropics, snow-covered boreal forests gird the Arctic, sporadic woodlands stretch across arid grasslands and half-submerged willows filter swamps in valley bottoms. Nothing is more typical of the land surface of our planet. Most of its biomass is trees. More than half those trees, and two-thirds of their carbon content, are in the tropics. But the largest single forest is stretched across eleven time zones of the Russian far north, which contains a quarter of all the world's trees.

Trees don't just dominate our living world; they made it. Before trees, some 300 million years ago, the continents were mostly hot, arid and lifeless. The atmosphere was very different. "Carbon dioxide levels were ten times higher, and temperatures ten degrees warmer," says Claire Belcher of the University of Exeter. "Oxygen levels were at half today's level." There was little soil. Fierce winds whistled across bare rock. But trees created a world in which they, and we, could prosper. "They transformed a barren planet and turned the world green," says David Beerling of the University of Sheffield.[2]

In the early days, they began by colonizing the only wet places, on coasts. They transpired moisture into the air, thus recycling rainfall from sea breezes to generate more rain in formerly arid regions inland. That allowed new trees to extend their domain, transforming the atmosphere as they went. As they grew in numbers, they also began to draw down carbon dioxide levels in the atmosphere, cooling the world and often extending still further the places where trees could prosper.

They congregate in vast numbers, remaking their local environments: for instance, by linking their foliage to create closed forest canopies that shade and cocoon the land below. They work together below ground too, being largely responsible for creating the soil in which they grow. Down there in the soil, they connect up once again in a kind of underground canopy. In recent years, it has become clear that the finest threads of tree roots and the fungi that live on them mingle in amazingly sophisticated ways, sharing nutrients, redistributing carbon, and passing chemicals that signal events around them.

Much of this secret subterranean network was uncovered by Canadian ecologist Suzanne Simard of the University of British

Columbia, who found fungal links in the soils of local forests of Douglas fir and paper birch. Using radioactive isotopes, she tracked compounds to prove that the trees are using the fungal networks to share resources and information. They look out for each other.

It now seems that most natural forests have such supply systems. Many plants and trees depend on them for essentials, such as nitrogen, water and even carbon created by photosynthesis in other plants. The networks deliver to order, in response to chemical signals from the trees. The fungi will even grab carbon and other resources from a dying tree and redistribute it to those in need. Large older trees that Simard calls "mother trees" are the most connected, but young trees starting out in a forest tap into these fungal networks from an early age, and even trees of species not previously found in the forest get to join in.

Following the word wizardry of an editor writing a cover line for the journal *Nature*, where the findings were first published, scientists now like to call these subterranean information superhighways the "wood-wide web." And the rest of us are catching on. In his sci-fi movie *Avatar*, director James Cameron depicts a distant moon called Pandora, which has a biosphere made up of a glowing underground neural network.

A forest is not just a bunch of trees; it "is a cooperative system," Simard says. It "behaves as though it's a single organism," controlling the forest environment and sharing resources among its inhabitants for what appears to be—and may actually be—a common purpose.[3] Some scientists find such language off-putting, unscientific, and decidedly un-Darwinian. How does "survival of the fittest," the cornerstone of evolution, find a place in this cooperative world? The fungi are not being altruistic, however. They are looking out for themselves, extracting a tithe for their services in the form of sugars made by the trees. And their control over the underground exchanges allows them to favor particular trees that help them most.

THAT'S WHAT WHOLE FORESTS can achieve. Now let's check what happens under the hood of each tree. Consider not the superorganism, but the tiniest and most vital feature of trees, the one through which they impose most of their power over the planet. These are stomata—the microscopic pores on leaves that take in carbon dioxide from the air and release oxygen and water. "Life on Earth depends in no small part on stomata," said the late Fred Sack of the University of British Columbia.[4] The pores take from the air the feedstock necessary for photosynthesis, the most important biological process on the planet, because it makes plants. We, like all other animals, get our food ultimately from plant matter. Plants, on the other hand, make theirs from the air by photosynthesis, a chemical process that uses energy from the sun to combine carbon dioxide from the air with water drawn up from the tree's roots. This creates glucose from which plant cells form. All this happens in leaves, where the stomata take in the carbon dioxide and then release the main waste products from photosynthesis: oxygen and excess water.

A single leaf can carry more than a million stomata. A single tree may have hundreds of thousands of leaves, so hundreds of billions of stomata. A big rainforest like the Amazon can have hundreds of billions of trees. That makes a lot of stomata. Ten to the power of twenty-one, perhaps. Stomata don't just open their pores to the air, Sack said. They also act as valves, regulating the inflow of carbon dioxide and outflow of water in ways that allow plants to optimize the use of both. They can shut down during droughts, for instance, while at other times they stimulate the tree to pump more moisture from the soil to boost photosynthesis.

By regulating this flow of oxygen and water for the tree, they also regulate them for the planet. I won't dwell here on the issue of oxygen. (Trees absorb as well as release oxygen. Things are in balance. It is a bit of a myth that the Amazon produces a fifth of the world's oxygen. It does, but it also absorbs just as much.) But the process by which trees create water is fascinating. Stomata release water into the air in the form of water vapor. This transpiration is what moistens the atmosphere, recycling the rain to keep the world wet. It happens on a stupendous scale, with stupendous consequences. The planet's three trillion trees release an estimated 14,400 cubic miles of water a year. Forming clouds as it blows downwind, this moisture is responsible for at least half of all the rain and snow that falls on land. In continental interiors that figure rises to more than ninety percent.

On the face of it, this is odd behavior. Pumping up moisture from the soil through the trunk to the leaves takes a lot of energy. It is vital for photosynthesis, but more than ninety percent of the water that reaches the leaves remains unused. It is transpired into the air through the stomata. Why does the tree bother? Certainly, the water cools the tree, but that could be achieved in other ways. There seems, instead, to be a collective purpose. Trees release moisture to make a world fit for more trees.

This role was laid bare a few years ago by Dominick Spracklen of the University of Leeds. He plotted data from thousands of rain gauges across the tropics and then looked at where the winds that brought the rain had come from during the previous ten days. His findings were dramatic.[5] From the Amazon to the Congo basin and Borneo, the story was the same. Air coming from forested areas delivered more than twice as much rain as air from deforested areas. Forests make rain; taking them away creates if not deserts, then certainly aridity.

The idea that trees increase rainfall is not new, of course. It goes back at least to Pliny the Elder's *Natural History* in Roman times. Most forest communities I have visited believe it implicitly. Christopher Columbus noted after crossing the Atlantic how it rained daily on forested Jamaica, yet only rarely on deforested islands he passed along the way, such as the Canaries and Azores.

Much later, British colonialists occasionally planted trees to improve the local climate. My favorite example of this deliberate tweaking of atmospheric conditions was the brainchild of Joseph Hooker, a botanist friend of Charles Darwin and future head of Kew Gardens in London. In 1843, he visited one of Britain's smallest and most remote outposts, Ascension Island. The arid extinct volcano, stranded in the middle of the South Atlantic more than seven hundred miles from anywhere, was entirely treeless—apart from a few Norfolk pine trees planted by the naval base there, in case they were needed to repair ships' masts. Water was in short supply. So Hooker came up with the idea of growing a forest on the volcano. "The fall of rain will be directly increased," he predicted, because foliage would scavenge moisture from passing clouds.[6]

The island was a refueling point for British shipping. So in response to Hooker's request, the Admiralty, the government department responsible for the navy, instructed sailors coming home from the far-flung colonies to bring trees and plant them on the mountain's slopes. Which they did. Two decades and some five thousand trees later, the Admiralty reported with satisfaction that the island "now possessed thickets of upwards of forty kinds of trees, besides numerous shrubs." As a result, it said, rains were improved and "the water supply is now excellent." The barren volcano was soon renamed Green Mountain, a name it retains to this day.

When I visited the island, the trees were still there too, many naturally reproducing. As I climbed Green Mountain, I walked through South African yews, Bermuda cedars, Persian lilacs, Brazilian guava trees, Chinese ginger, New Zealand flax, taro from Madeira, European blackberries, Japanese cherry trees and screw pines from the Pacific. I also saw monkey puzzle trees, jacarandas, junipers, bananas, vines, palm trees and Madagascan periwinkles. The summit was topped off with several acres of bamboo that rattled in the wind.

The island now has more than three hundred tree species, according to botanist David Wilkinson, currently at the University of Lincoln.[7] Ecologically, the accidental forest is fascinating. It is made up of trees delivered from all over the world and yet it appears to function as a single ecosystem. Insects that hitched a ride with the trees move indiscriminately from one continent's vegetation to another. This cacophony of trees had changed the climate and I was curious to find out what this meant for the local environment. Stedson Stroud, the island's conservation officer and my guide up the mountain, said that while the plains below were drier than before, the mountain was wetter (just as Hooker had predicted).[8] As we stood on the summit, the cool dampness of the mountain air contrasted with the heat of the plain. When noon approached, a lone cloud formed above us and then descended, shrouding the mountain (and us) in mist. A drizzle of rain followed.

THOUGH ROOTED IN OBSERVATION, the notion that forests generated rain fell out of scientific fashion in the nineteenth century. The new wisdom was that "rain follows the plow." In other words, that the best way of boosting rains was to get rid of the trees. This catchy phrase was coined by Charles Wilber, an American land speculator and part-time journalist.[9] It helped make him rich, as he sold off real estate to people seeking their fortune on newly deforested lands. His slogan stuck. It became the received wisdom, repeatedly cited in meteorology textbooks.

Received wisdom is a dangerous thing, however. Wilber's term had long puzzled Dominick Spracklen. He was troubled by the contradiction between the meteorological convention and what local people told him in the tropics about forests bringing rain. "I thought: let's see if the local stories are true," he told me. "And they were." It was the meteorologists who were wrong.

Meteorologists have always underestimated trees. They have long held that most rain falling on land begins as moisture evaporating from the oceans. Again, it's in all the textbooks. This supposition is true, up to a point. Yes, oceanic moisture keeps coastal regions wet. But that moisture usually rains out within a few hundred miles of the coast. Further inland, the sea breezes are dry. Unless, as Spracklen showed, there are forests to transpire moisture back into the air. The result is the difference between West Africa and the Amazon. Despite a strong sea breeze, known as the West African monsoon, precipitation decreases fast as you go inland from the West African coast. Eventually you reach the Sahara Desert. Make the same journey inland across the Amazon basin, however, and it can get wetter as you go.

It used to be presumed that rainforests had high rainfall because they grew in wet parts of the world. Now, thanks to Spracklen, we know that they often make their own rainfall. As do forests around the world. Take southwestern Australia, where the interior has seen a twenty-percent decline in rainfall since the 1950s, even though there has been no change on the coast. It might be a quirk of global climate change. But why would the interior show a different trend from the coast just a few hundred miles away? They get the same wind, after all.

Hydrologists Mark Andrich and Jörg Imberger of the University of Western Australia figure it was coastal trees that kept the interior wet. They say that on the southwest corner of Australia, an area of trees along the coast almost the size of England—including many majestic karri hardwood trees—have been clear-felled in the past half century. They have been replaced by wheat fields and the spreading suburbs of the state capital, Perth.[10] That coastal felling has dramatically reduced the proportion of coastal rains that get recycled. Without transpiring trees, the winds blowing inland are drier than before, so there is less rainfall away from the coast. As a result, crops have failed and reservoirs have emptied.

John Boland of the University of South Australia told me that the Eyre Peninsula in South Australia has also seen a suspicious decline in rainfall, following extensive local bush clearance since 1950. The good news is that it doesn't have to be permanent. The only exception to the decline has been an area known as the Monarto plateau, southeast of the state capital, Adelaide, where tree planting was carried out in the 1970s. The trees were intended to be part of a new "green" city. The city was never built, but the trees remained. Rainfall in neighboring towns such as Murray Bridge, where Boland grows fruit trees, is up by a third.[11] Were other factors at work? Perhaps, but it does all fit the pattern first identified by Spracklen.

Moisture recycling doesn't only keep the air wet. It keeps it cool too. When we step into a forest, we step into shade, and might assume that the air is cool simply because there is no direct sunlight. But it is much more than that. Forests are also great air conditioners. They do this because the transpiration of moisture from quintillions of stomata requires energy. Lots of it. That cools the air around. A single tree transpiring twenty-six gallons of water a day has a cooling power equivalent to two household air-conditioning units.[12] Multiply that by hundreds of billions of trees and you have a lot of cooling. On the edge of the Amazon, temperatures are five degrees Celsius (9°F) cooler in the forest than on nearby farms.[13] On the giant forested island of Sumatra in Indonesia, the air around the forests can be up to ten degrees Celsius (18°F) cooler than in adjacent palm oil plantations.[14]

Such effects can be local. In cities from New York to Hong Kong and London, areas with trees are on average more than a degree Celsius (about 2°F) cooler than those without. They can also influence temperatures over long distances. Researchers examining temperature patterns in the Amazon have found that deforestation as much as thirty miles away can cause measurable increases in maximum daytime temperatures compared to intact forests.[15] Carlos Nobre of the National Institute for Space Research, Brazil's top climatologist and a veteran of the UN's Intergovernmental Panel on Climate Change, calculates that rainforests keep the entire tropics more than a degree Celsius cooler than they would be without trees.

In temperate regions, transpiration rates are lower, so the difference is about half as much. Despite this, studies have found that forested areas have fewer hot summer droughts than areas without trees.[16] In North Carolina, forests deliver two to three degrees Celsius (around 4–5°F) of cooling during summer days.[17]

⸙

THESE COOLING EFFECTS COMPLEMENT and intensify the better-known global cooling effect of forests from storing carbon in their trunks, foliage, roots and soils, preventing it from entering the atmosphere as the warming gas carbon dioxide. Half of the biomass of trees is made up of carbon. Despite humanity's Herculean efforts to deforest the planet over the past century or so, there are still around three trillion trees. They contain as much carbon as mankind has deposited into the atmosphere since the start of the Industrial Revolution, or 130 tons for every human alive today.[18] A tree's relationship to carbon changes through its life cycle. Growing trees absorb carbon. Dying trees release it, as their biomass rots. Mature forests may be in a carbon balance, with losses counteracting gains.

Tropical deforestation is contributing to as much as fifteen percent of human-caused carbon dioxide emissions. Trees are fighting back, however, thanks to a phenomenon known as global greening. Besides warming the planet, rising carbon dioxide levels in the air also allow trees to photosynthesize more easily. Since around 1995, trees and other vegetation have been putting on a growth spurt across much of the planet. Satellite images have shown many places literally becoming greener. More growth means the trees take up more carbon dioxide, adding to their carbon stocks. This doesn't end the buildup of the gas in the air, or halt global warming, but it does moderate it.

While global greening is visible from above, it is hard to measure the carbon accumulation in the field. One researcher who has done so is Lan Qie, now of the University of Lincoln. In 2015, she traveled to Borneo to meet forest elders who back in 1958 had been involved, as young men, in a study to measure the growth rate of their local trees. She had records of the results and asked the elders to take her to see how the trees were doing. She found that every acre of intact forest had added more than eight tons to its carbon store over the past half century.[19] African forests have typically added nearly ten tons per acre, says her former colleague Simon Lewis of the University of Leeds. The Amazon too, despite degradation in many areas, is probably still an overall accumulator of carbon.[20]

One study calculated that this carbon capture by the world's trees could have shaved a quarter-degree Celsius off global warming since the early 1980s.[21] We should not feel complacent. Global greening is clearly not enough to stop global warming. It may also turn out to be temporary if, as many researchers believe, the end result of all that extra carbon dioxide is that trees "grow fast but die young."[22] When they die they will give up all that extra carbon.

For now, the greening of trees is doing us a good turn. But we should never forget that trees play by their own rules. They will adapt to local conditions and attempt to control their local climate for their own local needs rather than ours. In many of the coldest parts of the world, that means they are warming the air. Their slow growth means they absorb very little carbon dioxide compared to forests in warmer climates. While their transpiration does cause local cooling, it is mostly exceeded by the warming they cause through another effect entirely. Dark forest canopies stretching for thousands of miles across Canada and Siberia make the Earth's surface less reflective—much less so than the white of the snow they often replace.

The technical term for reflectivity is albedo. It is from the Latin word for white, which also gives us albino. Scientists measure albedo on a scale from one to zero. One is a perfect white surface that bounces most solar radiation right back into space.

Zero applies to a dark surface that absorbs all the incoming solar radiation, which it then usually radiates back into the surrounding air in the form of heat. White surfaces keep things cool; dark surfaces cause warming. Fresh snow has an albedo between 0.8 and 0.9, desert sand is around 0.4, grassland and crops typically 0.25, a continuous canopy of broadleaf trees is as low as 0.15 and a canopy of conifers or dark ocean waters gets down to 0.08.

The great forests of the far north are mostly dark conifers. In daylight hours, their canopy absorbs the incoming solar energy and heats the air around. Land which is not forested is often covered with snow, reflecting almost every scrap of energy right back into space. Of course, it is dark a lot in winter. When the sun is up, however, the albedo contrast could hardly be greater. There is a tenfold difference. The trees here actually drive warming.

Now, every tree in every forest anywhere in the world has competing warming and cooling effects. The albedo of dark foliage warms the atmosphere, while transpiration and carbon storage cool it. The balance between them is different, depending on where you are. In the tropics, trees grow fast and sweat profusely, so the cooling dominates. In cold climes, where trees grow only slowly and transpiration is low, but the albedo contrast is very high, warming usually dominates.[23]

In parts of Siberia—and probably Canada too—the great expanses of conifers deliver five times more warming than cooling. This is great for them, the equivalent of a warm blanket during the endless cold nights. It is another example of the ability of forests to control their own local climate for their own benefit. However, in an era of global heating, it is not so handy for forests elsewhere—or for people. By soaking up the sun, these trees are heating the planet.

There is something else too. While the effects of changing albedo are instant, those of both deforestation and forest regrowth are slower. They play out in atmospheric carbon concentrations over decades, as forests grow, or cut timber decays and soils degrade.

Even where growing trees ultimately have an overall cooling effect, this may take decades to emerge. This means deforestation can cause cooling, at least in the short term. So concluded a 2019 report from the UN's Intergovernmental Panel on Climate Change. It found that historical forest losses around the world may so far have caused more cooling than warming.[24]

That applies to North America too, according to Christopher Williams of Clark University.[25] He carried out a detailed study of the competing cooling and warming influences of forest change on the coterminous US (that is, ignoring Alaska and Hawaii). He found that for large parts of the country, the cooling albedo effect from deforestation remains dominant to this day. In the coming decades, the balance will change east of the Mississippi, but not in the Rocky Mountains. There, albedo cooling will continue to dominate, meaning that if trees are planted in the future they would cause local warming.

It is a troubling and counterintuitive finding. Away from the far north, the effects are not huge, but "expanding forest cover in these regions to combat climate change is likely to be counterproductive," concluded Williams. That doesn't mean we should refrain from reforesting. More trees will protect biodiversity and soil, and, as we will see, they generate rainfall, and may make winds and even cool their immediate environment. But we should be wary of the simple presumption that they will also cool the planet.

## — 2 —

# FLYING RIVERS

## *Chasing the Rain and Mapping a New Hydrology*

GERARD MOSS IS a bush pilot in the swashbuckling tradition. Born in Britain and raised in Switzerland, he had flown twice around the world in his single-engine plane before he set out on a new mission—to track rain clouds across the Amazon in his adopted home of Brazil. "My goal was to show Brazilians that they are sitting on a gold mine in terms of water, not just in the Amazon River but in the clouds above the rainforest," he told me as we toured an exhibition of his exploits in Lausanne, Switzerland, at the end of 2019. It wasn't only Brazilians he transfixed with his voyages, however. He helped to change global perceptions of why it rains across the world.

The idea for his flights came in 2006, when he met climate scientist Antonio Nobre, the younger, hipper brother of Carlos, then at the National Institute of Amazonian Research in Manaus. Moss and Nobre were both attending an environment conference, on a boat sailing up the Amazon. Nobre was intrigued by the theory circulating among climate scientists that the Amazon rainforest was South America's biggest rainmaker: in other words, that most of the clouds delivering rain across the continent contained moisture that had been taken up and recycled back into the air several times by the forest's trees. He believed, moreover, that this moisture was carried by a concentrated flow of wind recently identified by meteorologists as the South America low-level jet, which blew across the Amazon at the speed of a racing bike. He called the moisture-laden jet a "river of vapor," not unlike the giant Amazon River on the ground.

Between them, the two men cooked up the idea of testing the theory by having Moss cruise the clouds above the rainforest. Nobre told me, "I said to Gerard, 'You are a pilot, you could follow the rivers of vapor.'" Moss told me, "I had already spent time flying over the Amazon. Looking at clouds coming out of the forest canopy after the rain had got me really excited. So I said to Antonio, 'Let's go flying and get the proof. Let's do it.'"

Back then, Moss's aviator exploits had made him a minor celebrity. The former Rio de Janeiro playboy had a regular TV show on Sunday nights. So he set about tapping his contacts for cash to equip his plane for collecting water vapor. The Brazilian oil giant Petrobras bought in. Nobre assembled a science team to analyze his samples and then they were ready to go. At a meeting with the pair shortly before the first flight, meteorologist José Marengo, then at the National Institute for Space Research, coined the term "flying rivers." Now they had a catchphrase too. Moss was soon airborne.

In tracking and sampling the waters of the Amazon flying river, his thirty-two-year-old plane clocked up around seven hundred hours. "We flew about thirty feet above the canopy, so we could guarantee the water we were sampling came from the trees," he said. "At times we'd be flying in blue skies over the forest and suddenly out of nowhere the humidity would shoot up for ten or fifteen minutes and then go back down again. We were going in and out of the flying river."

Slowly, he built up a picture of the water flows as proposed by Nobre. "We could see that the moisture was initially coming in off the Atlantic, carried on sea breezes from the northeast. That moisture then rained out and was transpired back into the air every hundred miles or so." The river in the sky was temperamental. It came and went. Sometimes there was a fast, concentrated flow; sometimes it spread or meandered. It could grow to two miles high and hundreds of miles wide.

Sitting behind Moss in his plane all the way was an assistant operating a small pump to collect air samples. The assistant cooled the air with liquid nitrogen, to condense out droplets of water. Back on the ground, Moss took the samples to the University of São Paulo's Isotopic Ecology Laboratory for analysis. Eventually, the lab analyzed droplets from nearly a thousand samples, looking for chemical fingerprints that identified the droplets' origins, in particular to find out if they had evaporated from the ocean or transpired from the forest. Their analysis relied on the fact that water molecules from vegetation contain more atoms of deuterium (also called heavy hydrogen) than molecules from the ocean. It showed that something like half the moisture in the flying river, and therefore half the rain falling over the Amazon, had transpired from the rainforest. They had demonstrated to the world that the Amazon rainforest was in large measure making its own rainfall.

Initially Moss flew into and out of these "rivers." It was exciting: a bumpy ride. What he really wanted to do, though, was follow the flying river right across the rainforest. After several failed attempts, he managed to follow the river in full flood for eight days. From Belém, a city on the northeast edge of the rainforest close to the Atlantic Ocean, the river tracked west across the full width of the Amazon towards the Andes Mountains, then turned south and finally east towards São Paulo, the biggest city in the Americas.

This adventure captured the imagination of the media—the human story gave them a way of bringing the science to life. "I told the press that I had arrived that day in the same air that I had left Belém in, and that when it rained that evening the rain had for sure come from the rainforest," he remembered as we looked at some of the headlines on display at his exhibition. "The media loved it. I had a message, with the data to back it up. I could tell them that if we lost the rainforest, São Paulo would become a desert."

The message was heard loud and clear in São Paulo—at a time when the government of President Lula was cracking down on deforestation. It also spread to the jungle, where the governor of Amazonas phoned Moss to ask him how much all the water was worth, so he could send a bill to the southern farming states that relied on it to water their crops.

Moss's findings were, strictly speaking, not a scientific revelation. Nobre had been on the case for a while. Way back in 1984, Enéas Salati, the head of the National Institute of Amazonian Research, had said, "Clouds seem to arise almost from the forests," and warned that widespread deforestation would "set in motion a chain of events... There will be a reduction in the amount of water recycled to the atmosphere, and hence diminishing precipitation."[1]

The data Moss produced were compelling, however. They showed, for instance, that the flying river he followed from Belém to São Paulo was transporting more than 100,000 cubic feet of water a second—enough in a day to supply the twenty million inhabitants of São Paulo for almost four months.[2] The rainfall in the Amazon was being recycled not once but typically five or six times on its journey across the rainforest. More important, perhaps, the story of his flights was a brilliant and very public confirmation of the new science. He had flown across the Amazon, smelled its breath and taken the droplets back to the lab. "Scientists could say we were ground-truthing what they already believed. But for the public it made all the difference," Moss said over a final glass of wine at his exhibition launch in Lausanne, where he has a second home. "By understanding my journey, they could understand that the Amazon was their lifeblood, even for those living in the city." Recognizing Moss's ability to use this story to deliver a clear scientific message, the then Brazilian environment minister, Marina Silva, sent him to the UN climate negotiations in Copenhagen in 2009. "Brazil was proud to show it had flying rivers," Moss told me.

The Amazon flying river is the trunk water main of South America, delivering rain over half the continent. Its moisture maintains the clouds over Lou Jost's orchid-covered ridges in Ecuador, for instance. As Partha Dasgupta, an environmental economist at the University of Cambridge, puts it, "Every country in South America benefits from the Amazon's moisture, except for Chile."[3] Remarkably, thanks to the power of transpiration in the forest below, as the river gets further from the ocean, it delivers more rain. Near the ocean at Santarém, rainfall is eighty-five inches a year; by Tefé in the middle of Amazonas it is ninety-six inches. The phenomenon is even more marked at the river port of Leticia. On the border between Colombia, Peru and Brazil, in the far west of the Amazon basin, Leticia is cut off from Pacific Ocean breezes by the Andes Mountains and receives its winds from the east. Without trees along the way, it would be bone dry. With the winds passing over trees, however, it is still jungle wet. Leticia receives 130 inches of rain a year.

After reaching the Andes, the Amazon flying river is pushed south by the mountain range. Still saturated, it travels the western flank of the Amazon basin and on over southern Brazil, Bolivia, Paraguay and Uruguay to the Argentine pampas. Most of this southern leg of the journey is over the basin of the La Plata River, which as a result gets seventy percent of its water from the Amazon flying river. Marengo, now head of the National Center for Monitoring and Early Warning of Natural Disasters, says even partial Amazon deforestation would reduce dry-season rainfall in the basin by more than a fifth. It would empty reservoirs, dry up soybean fields and cattle pastures, and turn the great Pantanal wetland into a dust bowl. It would be like "cutting the tributaries of a real river," he told me. Only by keeping the trees can we keep the moisture and the flying rivers.

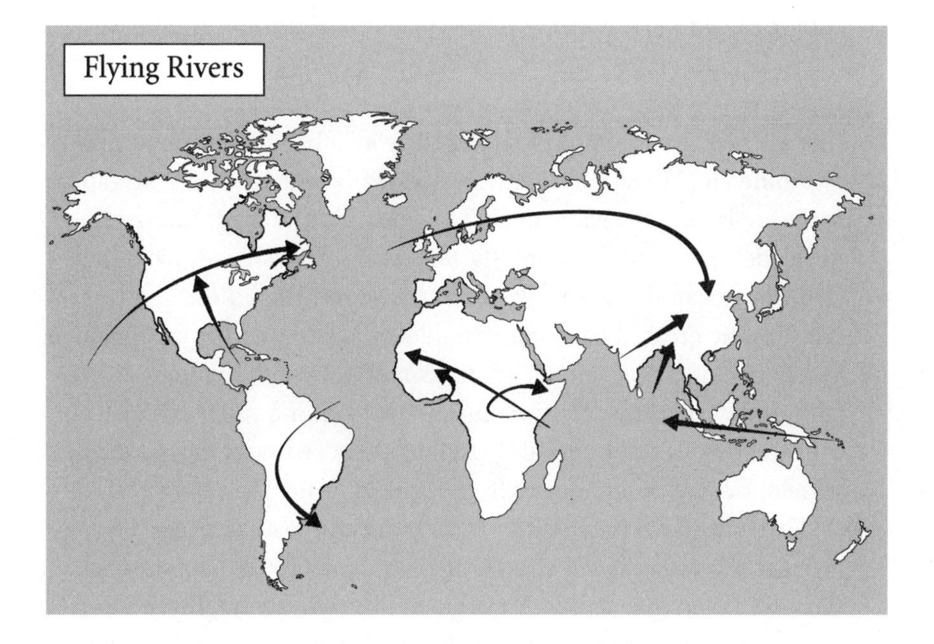


IS THERE ANYTHING SPECIAL about the Amazon flying river? Do rivers flow through the sky elsewhere? The sheer size of the Amazon rainforest gives its flying river a head start. But it turns out there are indeed others, many with equal importance for spreading rainfall generated by trees to continental interiors that would otherwise be arid.

Back in the 1990s, as Brazilian climate researchers were developing their ideas about Amazon water recycling, Hubert Savenije, a hydrologist at the Delft University of Technology in the Netherlands, began studying moisture recycling in West Africa. He found that as you moved inland, the proportion of rainfall that came from forests grew, reaching ninety percent in the interior.[4] There was a coastal wind for part of the year, called the West African monsoon. However, because there were so few forests, it soon dried out over countries like Mali, Niger and Burkina Faso. Eventually, there was the Sahara Desert.

Savenije set one of his bright young students, Ruud van der Ent, the task of figuring out if there might be a global dimension to this phenomenon. "I read the literature and realized there was a big gap in the analysis of the water cycle," Van der Ent told me. The conventional idea that rainfall was mostly generated by moisture evaporating from the ocean did not fit the data. He soon realized the gap was trees. So he created a computer model of air flow across the globe that fit real data on evaporation, transpiration, winds, temperatures and rainfall patterns. His findings, published in a series of papers over the past decade, reveal that globally more than forty percent of the moisture in rain comes from the land rather than the ocean.[5]

He showed me the maps he had drawn of moisture flows. They revealed rivers in the sky stretching out from all the major forests of the world. Along with the Amazon and its flying river, there were others delivering moisture across North America, through the forests of Southeast Asia, and from the Congo rainforest to both the Sahel and Ethiopia. The longest of these flying rivers swept west to east for four thousand miles across the Eurasian land mass. The further east it got, the more of its moisture had been recycled by forests. In northern China, eighty percent of the rain was generated by the forests of Scandinavia and Siberia, a journey involving several stages and taking six months or more.

"The China finding was among my first, and it was a real eye-opener," Van der Ent told me. "It contradicted previous knowledge that you learn in high school. China is next to an ocean, the Pacific, yet most of its rainfall is moisture recycled from land far to the west."

Real-world data support the printouts from his model. Just as the Amazon flying river can deliver more rain inland than at the coast, so do the forests of the Congo basin. At coast cities such as Pointe-Noire, they get around forty inches of rain each year, whereas deep inland at Kisangani, it is more than sixty inches.

The same pattern holds when traveling inland from the Arctic across the great northern forests. The recycled moisture fills great Siberian rivers such as the Ob, Yenisei and Lena. A traveler moving upriver will discover precipitation roughly doubling every twelve hundred miles. The same is true on the Mackenzie River in northern Canada. A quick check of published climate records shows that annual precipitation increases from six inches where the Mackenzie enters the Arctic Ocean to twelve inches at Fort McPherson in the northern Northwest Territories, thirteen inches at Wrigley, further south, and seventeen inches at Fort McMurray, an oil town on the Athabasca tributary amid the forests of northeast Alberta. There is no ocean to provide the moisture for this rain—but there are trees.

Where there are no interior forests, the story is very different: rainfall rapidly diminishes inland. Instead of rich forests there are deserts. In West Africa, where Savenije did his pioneering work, it drops from 180 inches in coastal Liberia to only two inches in Tamanrasset, an Algerian oasis twelve hundred miles inland. On the coast of northern Australia, Darwin gets seventy inches a year, but at Alice Springs, some thousand miles inland, they get only eleven inches.

Strange to say, when it was first published a decade ago, nobody took much notice of Van der Ent's modeling. The conveyor belts of moisture he identified were rarely acknowledged outside the Amazon. Most climate researchers saw the impacts of forests only in terms of carbon dioxide, or the local effects of transpiration. Now that perception is changing. "A new image of the global hydrological cycle is emerging," says hydrologist Lan Wang-Erlandsson of Stockholm University.

The implications are important. In much of the world, the loss of the moisture recycling from deforestation is a more imminent threat than global warming. Agricultural regions of China, the African Sahel and the Argentine pampas all depend on distant forests for their rains, says Patrick Keys, an atmospheric chemist at Colorado

State University.[6] Tropical forests are usually cleared to increase the land available for agriculture. By turning off the rain tap, deforestation may ultimately make farming untenable over even larger areas.

Forests provide water for distant cities too. Keys tracked the sources of the rain supplying twenty-nine global megacities. He found nineteen of them were highly dependent on faraway forest transpiration, including Karachi in Pakistan, Wuhan and Shanghai in China, and Delhi and Kolkata in India.[7] His top ten most forest-dependent cities have a combined population of 200 million people. "Even small changes in precipitation arising from upwind land-use change could have big impacts on the fragility of urban water supplies," he says.

In some of the world's largest cities the warning signs are already clear. Tree loss in India may explain why its giant eastern megacity of Chennai was hit by drought in 2019. For several weeks, the city's four main reservoirs dried up. Businesses closed and there were water riots. Tankers toured the streets with emergency supplies.[8] Those seeking a cause turned up a study that found transpiration from forests on the Western Ghats—a chain of mountains along the western coast of India—contributed up to fifty percent of the city's rain in dry years. Some thirty-five percent of those forests had been lost in the past century. In a paper published in 2016, Supantha Paul and Subimal Ghosh at the Indian Institute of Technology in Mumbai had predicted that this "recent deforestation will affect water availability" downwind.[9] It had only taken three years for their prediction to come true.

When I contacted them, the pair went further. It was not just Chennai. Forests determined the future wetness of Indian monsoon winds that bring eighty percent of the country's rain. They estimated that a quarter of all the rain falling in India during the second half of the monsoon season came from forests recycling the rains that fell during the first half.

The islands that make up Indonesia get much of their rain thanks to moisture evaporating from the surrounding seas. But Van der Ent estimates that Borneo, the largest island, receives around a quarter of its rainfall from its own transpiring trees. No wonder that a twenty-percent decline in rainfall across Borneo in the past half century has coincided with the loss of half its trees. This may also explain why the biggest decline in rainfall has been in the southeast corner of the island, where deforestation has been greatest.[10]

If more forests disappear, "it is likely that this tropical ecosystem will collapse," says Clive McAlpine, an ecologist at Australia's University of Queensland.[11] If Indonesia and other countries want to restore lost rains, a good bet would be to restore lost forests.

One morning in a café outside Berlin, I met Taiwanese geographer Wei Weng, from the nearby Potsdam Institute for Climate Impact Research. She had been working on what she called "smart reforestation." She showed me her plans for replanting seventeen million acres of deforested Bolivia to replenish the Amazon flying river and generate rain for Santa Cruz, its biggest city. Her modeling of how local forests recycle rainfall suggested that the replanting could deliver 500,000 acre-feet of extra rain annually (an acre-foot is the amount of water that would flood an acre of land to a foot in depth).[12] She even identified the best places for reinstating trees, in two departments on the border with Brazil: Pando and El Beni. Such specific plans might show undue confidence in the precision of the modeling, but the authorities in Santa Cruz were enthusiastic. Officials told me later they had identified some farms for a trial.

Santa Cruz may be only the first of many cities to consider such measures. Chennai could do with some trees back in the Western Ghats. In Brazil, São Paulo should be keen. The megacity almost ran out of water in early 2015, during a drought that the city leaders blamed on Amazon deforestation. The Cantareira reservoirs—the largest sources of water for the largest city in the western hemisphere—were down to their last six percent. Many taps ran dry.

ELSEWHERE, SOME SCIENTISTS are calling for a return to the methods used in the colonial era by Joseph Hooker to make rain on arid regions specifically by planting trees. But rather than trying to wet small islands, they have in mind the world's largest deserts. In 2018, Gil Yosef of the Weizmann Institute of Science in Israel made the case for planting forests on a massive scale in the dry desert margins of the Sahel in Africa and arid northern Australia.[13] He envisaged a forest the size of Mexico that would, he said, increase local transpiration enough to moisten and strengthen the West African monsoon, increasing rainfall in the Sahel and Sahara by almost twelve inches a year. In Niamey, the capital of Niger, that would be a rise of more than fifty percent. In drier places, the percentage increase would be even greater. More rain would encourage the new megaforest to expand without further human assistance.

In theory, it could return the Sahara to the climate it had six thousand years ago, when there were forests, lakes, giant rivers and lush wetlands, inhabited by herds of buffalo and crocodiles.[14] It could also, Yosef claimed, increase the take-up of carbon by the world's forests by a tenth. Such a project sounds crazy, but there is good scientific evidence that the Sahara dried out back then because some tipping point was passed, beyond which a failure of rains killed trees, reducing their ability to recycle moisture, drying the air further and causing more trees downwind to succumb. So why not put such runaway effects into reverse?

SOME CLIMATE MODELING SUGGESTS that forests also sustain weather patterns far beyond the paths of flying rivers, by influencing how energy moves around the atmosphere. Roni Avissar, a climatologist at the University of Miami, says that the Amazon rainforest maintains rainfall on the other side of the equator in the US Midwest grain belt and snowfall in the mountains of the Sierra Nevada.[15] The Sierra Nevada supplies both the cities of California and irrigated farms of the Central Valley, a major source of food across the country.[16] Avissar's number-crunching predicts that complete deforestation in Brazil would reduce precipitation by a fifth in faraway Oregon and Washington; cause a halving in the winter snowpack of the Sierra Nevada; severely reduce rainfall in the lower Midwest in spring and summer; and do likewise in the upper Midwest in winter and spring. These long-distance influences, often called teleconnections, are a ripple effect from the drying of air over the northern Amazon, projected around the globe by patterns of winds in the upper atmosphere known as Rossby waves.[17]

Avissar says this teleconnection effect from Amazon deforestation would be analogous to El Niño, the shift in the temperature of waters of the tropical Pacific Ocean. By changing rates of evaporation from the ocean, El Niño too flips rainfall patterns along Rossby waves and causes climatic upsets right around the world, with deserts drenched in rain and rainforests struck by drought. Of course, the time scales are very different. El Niño is temporary and lasts a few months, while deforestation takes decades and lasts much longer. That means we can see the effect of El Niño on distant rainfall, whereas the impact of losing the Amazon rainforest would be much slower. The impact could be just as big, however, and would be far more permanent. "The big point," says Avissar's co-author David Medvigy, now of the University of Notre Dame, "is that Amazon deforestation will not only affect the Amazon. It will hit the atmosphere and the atmosphere will carry those responses."

Keys and Wang-Erlandsson have encouraged governments to pay much more attention to the reliance of their citizens on rainfall generated in lands far beyond their borders. The arid African Sahel—which includes dirt-poor drought-prone countries such as Mali and Niger—gets ninety percent of its sparse rains from moisture generated outside the region. The main stem of the world's longest river, the Nile, comes out of the Ethiopian Highlands, which receive forty percent of their rains from the forests of the Congo basin.[18]

"We need an international hydrological agreement to maintain the forests of such vital source regions," says Wang-Erlandsson.

It is hard to overstress the dangers of deforestation for the flying rivers. The discovery that farmers and megacities depend on distant forests for their survival dramatically raises the stakes in the battle to protect and restore the world's forests. We can imagine future wars over flying rivers. Even so, the discovery of their importance also offers opportunities. Future hydro-diplomacy to maintain and augment flying rivers could make for interesting cross-border co-operation, with deals done between countries for the restoration of forests upwind to keep the rains falling and the taps flowing downwind. To protect water supplies for Beijing, China could ask Russia to extend its northern forests. The countries of the Congo basin might be paid to keep their trees as a means to ensure the flow of the Nile from Ethiopia through Sudan to Egypt. In this way, maintaining forests may also be a way of maintaining international peace. The playboy bush pilot certainly started something.

## —3—

# FORESTS' BREATH

## *Sniffing the Air and Shooting the Breeze*

I CAN GET VERTIGO looking out of an upstairs window. Even so, I was about to embark on the trip of a lifetime. Not just into a forest canopy, but climbing way above, till the trees looked tiny below. We set off before dawn. A three-hour drive out of Manaus, the jungle city in the heart of the Amazon rainforest. Then a couple of hours by boat down the Uatumã River, one of the Amazon's major tributaries, and another half hour on a low-loader truck through the forest to our camp. There, after stashing our backpacks on hammocks, my guide and I headed for the object of my pilgrimage. As the sun flashed through the forest canopy, and howler monkeys set up their cries, I caught sight of a metal tower soaring into the sky. At 1,066 feet, the Amazon Tall Tower Observatory (ATTO) is the tallest structure in Latin America and three feet higher than the Eiffel Tower in Paris—though, as I quickly noticed, far less robust. More like a radio mast, it has no elevators and is mostly just an open stairwell to a platform on the top.[1]

At the bottom, I listened intently to the safety briefing. There were fifteen hundred steps to go, rising ever higher above the forest canopy. We donned our safety harnesses attaching us to the structure, but were warned that because of a design problem, we would have to keep releasing and reattaching them at each of the hundred or so turns in the stairwell. Oh well, it was too late to back out now. I took a deep breath and headed up.

First, we were climbing through dark and steamy jungle, with the chatter and song of wildlife all around. As we rose out of the canopy a large bird flew past within inches of our faces. By now the weather was changing with almost every step. Temperatures peaked just under the canopy, but then started to drop. Humidity fell too. My somewhat shaky notes recorded that above five hundred feet, a stiff wind began to blow. I held on tighter, double-checked the safety harness at each turn and kept going.

The climb took almost an hour, with stops to admire the view as we went. Long before we reached the top, the trees below looked very small, like a mass of broccoli heads stretching unbroken to the horizon. There was a haze in the air too. This was mid-September in 2019, the year the Amazon burned. For hundreds of miles, the world's largest rainforest was pockmarked by fires. None were close, though on the horizon, a couple were visible, and on the breeze I could smell the burning.

I was with German researchers, who all the way up were checking their monitoring equipment. Strapped to the tower and hooked up by cable to data loggers on the ground, the instruments were recording the emissions of gases from the billions of trees stretching out below us. The forest contained forty percent of all the living biomass on the planet. The trees were breathing in carbon dioxide and exhaling water vapor and oxygen. The Germans' real interest, however, was the dozens of other chemicals also being released by the trees. Top of their list was a group of chemicals known as volatile organic compounds, or VOCs. These are thought to help create the clouds that recycle the moisture given off by the trees. Before my visit to the tower, I had heard the researchers present their findings at a conference in Manaus. Now I was heading into the clouds to hear more from the man who started it all.

As I stepped onto the uppermost platform, I was met by Meinrat Andreae, director of the Max Planck Institute for Chemistry in Mainz and the tower's mastermind. A quiet, self-contained man, he had traveled from Europe to attend the conference. He had gone ahead of our small group to climb the tower alone. No doubt he enjoyed being able to savor the forest, unencumbered by my questions. Now we stood together, sniffing the breeze, looking down on the trees and gazing at a forest extending for hundreds of miles in every direction. Finally, I had the chance to ask him how the tower came to be built.

Back in 2007, Andreae had proposed the structure to examine the rainforest's interactions with the atmosphere in more detail than before. It would be the first time anyone had installed permanent sensors in the sky high above a rainforest canopy. As we gazed out over the endless blanket of trees, he told me that he had originally hoped that this spot, a hundred miles north of Manaus, was remote enough to provide "a window on our planet's atmospheric chemistry before industrialization." It would be a place to analyze the breath of a vast and largely pristine forest.

He partly got his wish. During the wet season here, clean air from the Atlantic blows across five hundred miles of largely intact forest. The breeze is the Amazon flying river on its long journey across the continent. But we were sniffing something else that was decidedly less clean. Our climb took place at the end of the dry season, when the winds sometimes came from the south, from a region that Brazilians now call the "arc of deforestation." There was a strong stench of smoke from the fires raging through the forest. Besides a window on the pristine past, Andreae and his fellow researchers feared that the tower was giving them a window on the Amazon's dystopian future.

Since the tower was opened in 2015, its instruments have been sampling the air above the forest for carbon dioxide, sulfur compounds, pollen, VOCs, particulates, methane, nitrogen oxides and much else. The data are all logged in mobile labs at the tower's base, where researchers turn the raw numbers into other handy parameters, such as rates of photosynthesis in the trees and of the exchange of carbon and moisture with the atmosphere. All this information is building a unique picture of how the biggest and most complex ecosystem on Earth interacts with the atmosphere.

Some of the chemistry has been unexpected.[2] "There are reactions going on in the air that we haven't accounted for yet," Andreae told me. "That means there are probably tree emissions we are ignorant about too." That made him excited. It was why he built the tower, he said. In my experience, many scientists are only interested in doing research to confirm what they think they know. The really good ones, such as Andreae, revel in grasping at the unknown. He was trying to unpick the role of the complex and sometimes unknown VOCs that give the forest's breath its specific odor.

Most trees emit these gases. There are dozens of different ones. The main categories—or at least the best known, which may not be the same thing—are isoprenes and terpenes. Isoprenes are produced mostly by deciduous trees and make up about half the total globally. Conifers produce terpenes, which are responsible for the distinctive scent of pine. Among these, Andreae told me, his hunch was that fast-reacting sesquiterpenes could turn out to be the most important in the Amazon. These are hard to spot because they disappear very quickly. That means they are extremely reactive, which is what intrigues him. He wants to know what they react with, and what that reaction produces.

VOCs are emitted by leaf stomata, along with the water vapor and oxygen. They emerge as gases, but often quickly oxidize in the air, to form tiny particles known as aerosols that are vital in cloud formation. Clouds require moisture to form, of course. They also need what meteorologists call cloud condensation nuclei, tiny particles floating in the air. It is around these particles that water vapor condenses to form the droplets that eventually fall as rain. Many different particles in the air can do this, including salt from the oceans and air pollution, resulting in smogs. In remote forest areas, the main cloud condensation nuclei are the particles formed by VOCs. Investigating this process, and the complex chemistry involved, is the basic purpose of Andreae's tower. He wants, he told me as we began our descent of the tower, to see how it works both in clean "pre-industrial" air and in the smoky air caused by the millions of fires in the Amazon.

Will he be allowed to uncover these secrets of the forest's breath? As we drank beers in camp that evening, everyone gathered nervously around the TV to watch a speech being given at the UN's climate summit in New York by the Brazilian maverick president Jair Bolsonaro. He railed against foreign interference in the Amazon, denouncing what he saw as their "colonialist" aspirations to take control of the rainforest under the guise of conservation and science. With the howler monkeys cranking up the noise in the dark, the German scientists looked perplexed. Did Bolsonaro mean them? Their Brazilian hosts worried aloud that their European collaborators could soon be sent home.

As I write, a year on, foreign scientists are being kept away from the Amazon and the tower, but only by the coronavirus lockdown. They still look forward to returning—and to continuing to uncover the secrets of the Amazon's breath. But the threat to their work remains.

⁂

ALL FORESTS EMIT VOCS. In summer, the eastern United States is a hotspot. VOCs make the famous haze over the Smoky Mountains.[3] Globally, trees emit an estimated billion tons of VOCs a year. Divided among three trillion trees, that is about thirty-five ounces each. Trees expend a lot of energy making these complex chemicals, which have such a short life when they hit the air—about half an hour in the case of isoprenes. Biologists have long wondered why they bother. Theories abound. It seems that some of them repel pests and may attract pollinators. The hazes and clouds that they can create may allow trees to cool off in hot weather and may trigger rainfall. Some of the compounds are flammable and can act as fuel for forest fires during droughts—which can be handy in the many forests where fire is essential for tree reproduction.

Most of this sounds good and useful. It fits our image of forests as the environmental good guys—the pinnacle of nature's achievement. All of which may be true. But trees have their reasons for doing things that don't always meet our expectations. Sadly, anyone who points that out may find things getting rough, because the science can end up being intensely political.

As I mentioned, VOCs help create clouds because they get oxidized into tiny particles around which cloud droplets form. This oxidation happens when they react with a natural chemical in the atmosphere. Called hydroxyl, this chemical is the atmosphere's main cleansing agent. These days a lot of this hydroxyl is consumed in reactions with the virulent greenhouse gas methane. So now we have a problem. If there are lots of VOCs around, their reactions with hydroxyl use up so much of the stuff that there is less left over to neutralize methane. As a result, methane sticks around in the air for longer. Its concentrations build up, heating the air. VOCs may help make it rain, but by causing a buildup of methane, they also contribute to global warming.

So here is a question: Does that make VOCs atmospheric benefactors *and* villains? Could that make trees atmospheric villains too? Enter a young scientist at Yale called Nadine Unger. In 2014, she wrote an op-ed for the *New York Times* about her research into VOCs. It was published just as world leaders were gathering in New York for a climate summit, where they called for a global campaign to protect and restore forests as a means of reducing global warming. In the article, Unger wrote, "This conventional wisdom is wrong.

In reality... considering all the interactions, large-scale increases in forest cover can actually make global warming worse." Pouring fuel on the fire, her editors headlined her article "To Save the Planet, Don't Plant Trees."[4] Unger said her calculations showed that, by boosting methane levels in the air, forest-generated VOCs caused so much warming that they negated all the cooling benefits of trees—from transpiration, cloud making and absorption of carbon dioxide.[5] Since 1850, she said, deforestation had not been adding to global warming, as generally supposed, but had actually cooled the planet, probably by about a tenth of a degree. So planting trees could be counterproductive. It could warm the planet. As her op-ed told the world leaders then gathering in New York, "More funding for forestry might seem like a tempting easy win... but it's a bad bet."

The op-ed was met with a storm of abuse. Fellow climate and forest scientists were outraged. Most used intemperate language and gross simplification to accuse her of using intemperate language and gross simplification. "This opinion was wrong at so many levels that it is hard to cover them all here," wrote one. So hard, in fact, that he failed to address her central scientific point at all. For him and many others, it was the "message" more than the science that they objected to. They hated her for questioning an article of faith among most environmental scientists: the virtue of forests. She was branded a climate-change denier. She received death threats. Even on campus at Yale, some colleagues refused to speak to her.[6] She was forced into exile, which is where I met her five years later, living and working on the other side of the Atlantic at the University of Exeter.

She was bruised, but combative. "It is still very threatening when women speak out about science," she told me. Much of the criticism she received was misogynistic. (She had found that only a tenth of science-based op-eds in newspapers are by women.) "It was extremely distressing. It really harmed my career," she told me.

"But now I am in a much better place." She stood by her analysis and just wanted her fellow researchers to do the math too. If she was wrong, then so be it.

Some took up the challenge. "She got a bad rap. A lot of abuse," Dominick Spracklen of the University of Leeds told me. He didn't believe she was right, but as a challenger of received wisdom himself, he set his team to test her argument. The critical question was not whether VOCs used up hydroxyl and lengthened the atmospheric life of methane. They clearly did. What mattered was whether the resulting warming outweighed the cooling effect that VOCs caused by creating clouds. Spracklen's colleague Cat Scott redid Unger's calculations and concluded that she was probably wrong.[7] Scott felt that in most places at most times, trees were still a cooling influence. "It's still an open question, though," she told me.

There is a lot we don't know about the forests' breath and its impact on climate. So we need more consensus-busting scientists—people who are just amazed at the versatility and sheer cleverness of trees, and how they make their climate according to their needs, not ours. I came across quite a few of them in researching this book. A lot of them are women. Here is another.

SUNITHA PANGALA IS a British post-doc researcher. She made waves in 2017 after spending two months traveling the Amazon River in a small boat, measuring gases coming from the floating plants, the soils and the stems and leaves of almost 2,400 trees on thirteen floodplains. She was looking for methane. She found it in abundance. "The trees all emitted a lot. In flooded parts, the trees become a massive chimney for pumping methane into the air," she told me. She found hundreds of times more being emitted than previously recorded—several pounds a day from every acre of flooded forest. She calculated that the trees of the Amazon were releasing more methane than all the tundra ecosystems of the

Arctic together.[8] It now seems that most of the world's trees emit methane at least some of the time.

Forest scientists have long amused their students by cutting small holes in tree bark and setting fire to the gases hissing from the trunk. The first recorded analysis of these flammable gases was made in 1907, when Francis Bushong, a chemist at the University of Kansas, cut a campus cottonwood and found they were sixty-percent methane.[9] Nobody took much interest. Now, however, in an era of climate change, those blasts of methane are no longer just a field-trip novelty. Methane is the second most important greenhouse gas. Countries have to account for their emissions. There are targets for bringing them down.

Pangala had a tough job getting her measurements accepted. Other more senior (and more male) researchers had previously seen forests as absorbers of methane. They were reluctant to be called wrong. But her findings in the Amazon left no room for doubt, says Kristofer Covey of Skidmore College, New York, who had been measuring methane in US forests: "She blew the story open. It was very difficult to argue with." It explained too why satellites flying above the Amazon routinely measured twice as much methane coming out of the rainforest as researchers on the ground could account for. Now Pangala had uncovered what they were missing.

Researchers have long cataloged methane emissions from microbes in rotting landfills, rice paddies and the guts of cattle, and fossil methane leaking from coal mines and natural-gas pipelines. They have also long known about emissions from natural microbial activity—in wetlands, for instance. Methane bubbling off them used to be known as "marsh gas." But, until Pangala, trees had been largely left off the list. This was a big mistake. Trees, it turns out, make methane in their foliage, in their trunks and by feeding raw materials for methane generation to the microorganisms living among their roots. And they do it everywhere, not only in wetlands. Why? Is it just a waste product? Or does it serve some hidden purpose? Nobody yet knows. The mysteries of the forests' breath are yet to be fully unraveled. Another outstanding question is whether that breath is the driving force behind the world's winds.

⸙

EVERY SUMMER, as the days get longer, Russian nuclear physicist Anastassia Makarieva leaves her lab in Saint Petersburg to take her holiday in the vast conifer forests of northern Russia. "The forests are a big part of my inner life," she says. Over the past twenty-five years, these annual pilgrimages have become an increasing part of her professional life too. For as she has camped amid the spruce and pine along the shores of the White Sea, and kayaked down their sparkling rivers, taking notes on nature and the weather, she has pondered a theory developed with her mentor, traveling companion and colleague at the Petersburg Nuclear Physics Institute, the late Victor Gorshkov. It is a theory about how Russia's northern forests, the largest expanse of trees on Earth, generate the winds that regulate the climate of much of Asia—and how other forests on other continents may do the same.

As we have seen, the vast land mass of northern Eurasia, stretching from Scandinavia across Russia and through Mongolia to China, has the longest of the flying rivers. Winds that start in the Atlantic and blow east are kept wet by accumulating moisture transpired by the great northern forest, an area of trees even bigger than the Amazon. The rain clouds carried by the flying river keep the giant terrestrial rivers of eastern Siberia flowing, and water China's northern plain, the breadbasket of the most populous nation on Earth.[10]

Looking at this relationship between wind and rain, Makarieva and Gorshkov took a step back to ask: What makes the winds that carry that moisture? What drives the flying river eastwards? Their controversial answer is that the forests make the winds too. They called their giant global wind machine the biotic pump. After Unger and Pangala, Makarieva is my third maverick woman scientist—prepared to go where the boys fear to tread. I won't trouble you with the mathematics but here, in essence, is what the argument is about.

It has long been the received wisdom of climate scientists that winds are driven by differences in air temperature. Mostly that means differences between air over the land, which heats up strongly in summer but gets mighty cold in winter, and air over the ocean, which holds on to heat better and changes temperature with the seasons much less. In summer, hot air over the land rises and cooler ocean air moves in. In winter, the reverse happens. Everything from continental monsoons to local sea breezes is described in this way.

Makarieva sees this as a sideshow. She says most winds are actually driven not by temperature differences but by the transpiration of forests. Transpiration from tree leaves releases water into the air in the form of gaseous water vapor. This is buoyant and rises above the forest, where it cools and condenses into droplets of liquid water. That much is not controversial. But Makarieva points out that this liquid water takes up much less space in the air than the water vapor. The change, she says, creates a partial vacuum. To fill the vacuum, more moisture-laden air is sucked up from below. This rising wet air will itself be replaced by air moving in horizontally. And so a wind is created—a wind that will keep blowing as long as there is sufficient water vapor to fuel the process. In other words, as long as there are forests.

The idea sounds straightforward, but because the old idea that winds are driven by temperature differences has been entrenched for so long, this alternative explanation had barely been considered. "Nobody had looked at the pressure drop caused by water vapor turning to water," Makarieva told me when we met in Vienna, during a lecture tour she was giving across Europe to promote her ideas.

Of course, the conversion of each water molecule from gas to liquid has only a minute effect on air pressure. But in the huge volumes of water we know are transpired from quintillions of stomata in trillions of trees in giant forests covering hundreds of millions of acres, the effect is potentially extremely large. Forests, she says, "are not just the atmosphere's lungs, they are its beating heart, too. The biotic pump is the major driver of atmospheric circulation on Earth."[11]

The theory has faced a headwind of criticism from mainstream climate scientists. Partly this is cultural. Science, as I know from forty years of reporting, can be surprisingly tribal. Makarieva and Gorshkov have been outsiders: theoretical physicists in a world of climate science, Russians in a field dominated by Western scientists, and, in Makarieva's case, a woman too. The reluctance may also be because the theory's underpinnings lie in complex theoretical physics tied to fiendishly complicated mathematical calculations. This stuff is beyond me, but it also seems to make most climate scientists defensive. They can't get their heads around it. "It's just mathematics," one top climate modeler told me, as if this were grounds for dismissing the theory.

The calculations and conclusions have received support from some atmospheric physicists and been published in respected physics journals. Though it has been hard work. Their first paper, published in 2007, was ignored.[12] In 2010, they tried again in a big journal read by many atmospheric scientists, *Atmospheric Chemistry and Physics*. It took three years for the new paper, titled "Where Do Winds Come From?," to get into print.[13] The editors added an unusual rider, saying publication was "not as an endorsement" but "to promote continuation of the scientific dialogue on the controversial theory [that] may lead to disproof or validation." A decade on, sad to say, there has been neither validation nor disproof, but instead a standoff.

Veteran climatologist José Marengo told me in Brazil, "I think the pump exists, but how important it is we don't know. On the other hand, the Russians are the best theoreticians in the world, so we need proper field experiments to test it." Douglas Sheil, a forest ecologist at the Norwegian University of Life Sciences, said, "All I have learned so far suggests to me that the biotic pump is correct. Even if we thought the theory had only a small chance of being true, it would be profoundly important to know one way or the other." Disappointingly, when I tried in 2020 to get comments from top climate scientists responsible for some of the big climate models, most still refused to engage. Even those reluctant to dismiss it out of hand said that their explanations for the world's winds work well already. If it ain't broke, don't fix it.

Is that true? In the top journal *Science*, I found a review pointing out that the current climate models cannot accurately predict tropical rainfall. Even the best could explain only half the Amazon's rainfall. This is a big deal. Tropical rainfall is one of the most significant, most dynamic and most consequential forces in the entire climate system. The reviewers called the failure "a major roadblock to progress in climate science." Which makes it all the more surprising that the biotic pump remains sidelined.[14]

One of the unexplained features of the climate that the pump could resolve is what Brazilian climate scientist Antonio Nobre, a biotic-pump fan, calls the "cold Amazon paradox."[15] Remember that conventional theory holds that winds blow from cold areas to hotter areas. Well, the winds rushing off the Atlantic and across the Amazon don't seem to follow this convention. Instead, they keep blowing across the Amazon even during times of the year when the Amazon is cooler than the Atlantic. Nobre says the only explanation is that the winds are not responding to the temperature difference, but are being drawn in by the forest's biotic pump.

NOBODY HAS YET COME UP with a viable and agreed experiment to test the pump theory on a big scale in the real world. However, one man has tried to demonstrate on a small scale that condensation does create a wind. I have known Peter Bunyard, a mercurial environmental thinker and author, for many years. He first introduced me to the idea of flying rivers and their potential global impact on climate back in the 1990s. Recently, he has been busy rigging up a biotic pump in his back garden in Cornwall. I took the overnight train from London to watch a test run.

After breakfast, as the sun rose over the trees on a neighboring field, we headed for a garden shed where he had built a laboratory to show that condensation of water does indeed drive a flow of air. Inside the lab—amid a maze of cables and sensors—he had rigged up some copper cooling coils from a refrigerator. When switched on, they chilled the air enough to cause a few grams of water vapor to condense out in the humid morning air, forming water droplets. He was mimicking the condensation you would get as moisture from transpiring trees rose above the forest canopy. About six feet above the coils, Bunyard had built a wide doughnut-shaped chimney, at the top of which was an anemometer—a device used by meteorologists to measure wind speed.

He turned on the refrigerator coils and we watched. For a few seconds nothing happened, but then a thermometer near the coils registered that the temperature of the air was falling. As the cooling reached three degrees Celsius (about 5°F), we looked up and saw that the anemometer was registering air flow. It wasn't exactly a gale. The wind speed rose to 0.18 meters a second, or 0.4 miles per hour. It persisted. More condensation produced more air flow up the chimney. Then Bunyard turned off the coils, and as the air temperature rose again, the anemometer registered zero. "It is so clear," he said with a smile of triumph. "The rate of condensation determines the air flow."

Outside the shed, Bunyard opened the back of the enclosed chimney. On the floor was a tiny pool of water, "rain" from the few droplets of condensed moisture that had been enough to trigger the air flow. He mopped it up and weighed it: almost two ounces.

"I sometimes get hailstones," he said. His experiment looked a bit Rube Goldberg and his shed hardly reproduced lab conditions. At the start, he had used toilet paper hung in the chimney to detect air flow. Then he recruited a retired British government hydrologist, Martin Hodnett from the Centre for Ecology and Hydrology, to put in sensors and data-logging equipment. The data from repeated experiments are clear and precise—condensation and air flow are in lockstep—and have recently been published in peer-reviewed journals.[16]

Either this was a magician's trick or his experiment was doing what the Russians said it should. "Everyone said I wouldn't get an air flow, but I did," he told me. "The air circulating in the doughnut can only be explained by the pump. If it works with the crude sensing equipment I have here, that shows it is a robust response. I believe it shows the climatologists were wrong to deny the biotic pump."

Disappointingly—for Bunyard, for Makarieva and for science—nobody either for or against the biotic pump theory had taken the trouble to go and look at what he has been doing in his Cornish garden. "The honest response of climate modelers would be to try and replicate my findings in a controlled laboratory, and then model the phenomenon to see how important it is compared to other things going on," he said. That is how science is supposed to advance, with theories being tested and research replicated. "But they don't want to know."

Back in Saint Petersburg, Makarieva is confident that her ideas will eventually win out. She says the physics makes it unassailable. "There is a natural inertia in science," she says, adding with dark Russian humor a quip of the legendary German physicist Max Planck that science "advances one funeral at a time."

While she waited for the mainstream scientists to catch up (or die), she joined a public dialogue in Russia about revising its Forest Code, the basic forest law in the country. The code says that all Russian forests, outside a few strictly protected areas, are open to commercial exploitation. If loggers can get there, there is no legal ban on them cutting the timber—even in the great forests of the far north that she loves so much. But there was change afoot. "Some representatives of our forest department got impressed by the biotic pump and want to introduce a new category of climate-protection forests that should be exempted from any exploitation." It made a welcome change, she agreed, to be part of a new consensus rather than the perennial outsider. It may also be a sign that the importance of trees for climate is at last being understood in Russia, a country with more forests than any other.

If Russia may be recognizing the true importance of its forests for the planet and its life-support systems, what of Brazil? In 2020, it seemed to be heading towards a dangerous tipping point.

— 4 —

# IN TANGURO

## *Tipping Points in Soybean Fields Foreshadow Crisis in the Amazon*

A SIXTEEN-HOUR BUS RIDE through the cold night had taken me from Brasília, Brazil's bright modernist capital, past huge scrubby cattle pastures, fields of soybeans stretching to the horizon and occasional patches of woodland, into the agricultural heartland of Mato Grosso. At dawn I had disembarked at Canarana, the last town on the road to the rainforest.

Full of grain silos, bars and John Deere franchises, Canarana was founded in 1975 to service a private colonization project that has since become the heart of the largest soybean-growing region on Earth. The region now looks very different from the way it appeared to the Edwardian adventurers who came prospecting here for gold. Over breakfast, locals related how, a century before, what was then a river crossing had been visited by an eccentric British explorer called Colonel Percy Fawcett. From there, he had headed into the jungle with his teenage son, in search of the "Lost City of Z," a mark on an old map that was supposedly where the gold lay hidden. The pair were never seen again, though at least thirteen expeditions have since set out in search of them, in later times often accompanied by TV crews. Local tribal elders say that Fawcett senior was killed by their great-grandparents for slapping a child in Tanguro village. Which, coincidentally, was where I was heading.

There is no longer a village at Tanguro. Instead, two hours' drive from Canarana, down a maze of side roads where we passed giant trucks pulling trailers full of soybeans destined for China, I came to the gates of what has replaced it: Tanguro farm. Covering an area the size of fourteen Manhattans, it is one of the largest soybean farms in Mato Grosso, or anywhere on the planet. The land was originally deforested for cattle pasture in the 1980s, as part of the illegal razing of state forest lands that for two decades engulfed much of Mato Grosso. Then it was bought by Blairo Maggi, boss of the world's biggest soy conglomerate, the Amaggi Group.[1]

It may once have been a wild and lawless frontier, but today there is an aura of calm and high-tech professionalism at Tanguro. The farm's 350 workers are commanded by a cadre of managers wearing navy-blue monogrammed company uniforms. Anxious not to be seen as a deforester, the manager insisted, "We have enough land now. Our main goal is growing higher-profit crops on existing land." Alongside his soy and corn, he planned to plant cotton soon, because the price was high.

We moved on from the company's temperature-controlled offices. I had come to see a different kind of air conditioning. In partial compliance with the government's Forest Code, considerable parts of the farm were left in an unmanaged state of wildness, creating little pockets of forest. Beyond them, we came eventually to the front line, where the farm stopped and the rainforest proper started. On one side of the narrow dirt track was hot bare ground, waiting to be planted. On the other was cool, moist rainforest, tall trees stretching intact northwest for hundreds of miles through the Xingu Indigenous reserve. This was where straight lines, monocultures and combine harvesters came face to face with the most biodiverse ecosystem on Earth. This was where the trees themselves cooled the air.

My bus companion, and host on the farm, was forest scientist Michael Coe of the Woodwell Climate Research Center (formerly the Woods Hole Research Center) in Massachusetts. Since 2004, the tall New Englander has lived and worked in old cowhand accommodation on the farm along with other researchers from Woodwell and the Brasília-based Amazon Environmental Research Institute (IPAM).[2] They have been monitoring the boundary between farm and forest to see how the abrupt change of land use is altering the local climate and natural vegetation. It was a model for what was going on elsewhere in the Upper Xingu basin, and more widely across the Amazon and beyond across the tropics—"the perfect laboratory," as Coe said. "We have over a decade of data here. Nowhere else in the tropics has that."

The climate changes he had seen were devastating. It was becoming clear from the investigations at Tanguro that forests modify their climate on a scale rarely realized, cooling the air and making the rain on which they themselves depend. It was also apparent that deforestation along this boundary was destroying the forest's resilience—and in the process undermining the synergy between forests and climate. I had read academic papers about how models predicted the existence of dangerous tipping points in the climate system that could be triggered by forest loss. Here was one, probably the largest, said Coe. If the Amazon was going to die, it would begin here. His fear was that he was already watching it happen in real time.

Joining us at the forest edge was Coe's research colleague Divino Silvério. He was a local, the son of poor farmers who had come to Mato Grosso on one of the government's periodic programs to settle the rainforest. While working as a farmhand nearby, Silvério Jr. began volunteering as an assistant at the research station. He was smart and assiduous. He was now its chief on-the-ground researcher, gaining a doctorate and authoring a string of scientific papers with Coe and others in top journals.

Those papers told a story. More than a third of the Upper Xingu basin, an area the size of England, had been deforested. That included most of the area outside the densely forested Xingu Indigenous reserve that abutted the Tanguro farm. By reducing transpiration, deforestation had dramatically raised local temperatures. The air over the farm was on average five degrees Celsius (9°F) hotter than in the forested reserve on the other side of the fence: thirty-four rather than twenty-nine degrees (or 93°F rather than 84°F).[3] The difference rose to ten degrees Celsius (18°F) at the end of the dry season.

A typical Amazon tree sweats around 130 gallons a day. The physical process of releasing this moisture requires a lot of energy. Coe calculates that ten square feet of intact rainforest removes from the air around it the heat equivalent of about two sixty-watt light bulbs.[4] Across the Upper Xingu basin, as the trees have disappeared, annual transpiration has halved. It has become hotter and the annual dry season is almost a month longer. It barely rains for almost four months now. There are dust bowls, and the remaining trees find it ever harder to survive. The meteorological data suggest that the tipping point has arrived. Getting out of the car, Silvério said he would show me what that meant right there and then for the forest.

Along the track, some trees were marked with paint. Silvério explained that they were some of the 2,800 trees of 150 rainforest species whose health he had been tracking for the previous four years. In that time, some of the biggest had died. We clambered over their trunks, which had fallen across the path. In their place grew grasses and trees that were native to savannah regions far to the south, where they were adapted to longer dry seasons. His forest census was showing the beginning of the "savannization" of the rainforest, he said. The deforestation was changing the climate, and the changing climate was flipping the entire ecosystem from jungle to savannah.[5]

Back at the camp, Coe pointed to a large empty building. It was, he said, where the farm's managers had nurtured thousands of rainforest saplings for planting around the farm to comply with the rules set by a previous government for keeping trees on farms. But in the new savannah climate caused by the farm, the planted saplings had died. The project was abandoned. "It is too hot now," he said.

Once underway, savannization could happen fast here and would be hard to stop, Coe told me over a barbecue that evening. He was being careful with the flames. The longer dry season, higher temperatures and opening of the forest canopy to sunlight all encouraged wildfires, he said. Parts of the Amazon that once had fires every hundred years now burned every six years. When the fires burn the rainforest, there is no way back. Savannah species take over almost immediately. "Fire is nature's way of starting over," he said, stamping out the embers from the barbecue as we headed off to bed.

The next morning, we took a last tour around the farm. Just outside the farm gate, we saw the blackened remains of a surviving fragment of forest that had been engulfed in flames just a week before. Silvério said a local rancher who had been clearing scrub from his pasture claimed that the fire had got out of hand and crossed the road into the forest. Was it in truth a deliberate attempt to take another piece of forest land for cattle raising? I asked. Silvério shrugged. Out there on the edge of the rainforest nobody knew—or at least nobody was saying.

As yet there was nothing inevitable about the process. The tipping point was near, but big landowners like Maggi were no longer clearing forests. Jungle nature was hanging in there. The soybean fields were not as devoid of wildlife as I had expected. Dozens of tall flightless rhea birds wandered around, pecking for seeds in the fields. We saw tracks left by tapirs. They are reputedly rare and reclusive, but their droppings were everywhere. At the forest edge, there were armadillo burrows. To my huge delight, I even got to photograph a young jaguar that we saw sauntering down a track just fifty feet away from a field waiting to be planted with soybeans. If felt almost as if the forest's inhabitants were hunkering down, waiting for these crazy farmers to give up and go away.

They may not be alone in taking the long view. To the north of us, stretching for hundreds of miles, was the Indigenous reserve occupied by several Xingu tribes. Here the trees still stood tall and intact, thanks to the six thousand or so inhabitants who still live in large part from the bounty of the forest. Their good forest management, combined with the continued resilience of the forest itself, means that the reserve remained eighty-five-percent forested. The reserve was not unaffected by the clearances all around it. Savannization nibbled at its edges, but as long as the Indigenous communities remained, their forest stood a chance. If there was a Lost City of Z in there somewhere, it was best that it remained lost.

FOR THE NEXT LEG of my Brazilian journey, I flew south to the state of São Paulo and a city I had never heard of, São José dos Campos. The longtime center of Brazil's aerospace industry, it was now also home to climate scientists who analyze satellite images of the Amazon. I was keen to speak with Carlos Nobre, from the National Institute for Space Research (INPE), who is a fierce protagonist for saving the Amazon. I had arrived over the weekend and he invited me to Sunday lunch.

Nobre had first predicted the savannization of the Amazon back in 1990. He calculated that the tipping point would be reached when deforestation hit forty percent. Recently, he had revised that estimate. Taking account of background global warming, he figured a twenty-percent loss could be enough to trigger irreversible change.[6] The satellite images being analyzed down the road at INPE put the current loss at seventeen percent in the Brazilian Amazon, he said. So the ecological doomsday could be close. Recent droughts in 2005, 2010 and 2016 "represent the first flickers of this ecological tipping point," he said.

"When the dry season lasts longer than four months, tropical forest turns to savannah," he told me. I paused, mid-mouthful. That was exactly the state of things that Silvério and Coe had described around Tanguro farm. Once the four months were exceeded, Nobre said, things would go downhill fast. There would be repeated fires. Changing climate at the forest edge would penetrate further into the forest, setting off a wave of savannization that might not stop till the rainforest was gone. Once it passed the tipping point, it might not matter if humans were removing the trees or not. Nature would do the job on its own.

After my lunch with Nobre, I was in a somber mood when I arrived the next day at INPE. It is not often that you meet a scientist breathless with excitement about their new findings. But as I stood talking with another researcher by the elevator, we were interrupted by an atmospheric chemist called Luciana Gatti. She was rushing to tell her colleague the results of her latest analysis of carbon dioxide absorbed and released by the Amazon rainforest, which she had completed that morning. "We have hit the tipping point," she almost shouted, caught between the buzz of discovery and anguish at the potential consequences.

For a decade, Gatti's team had been sampling the air from sensors on aircraft flying over the rainforest. She had been collating recent results. They showed that, perhaps for the first time in thousands of years, a large part of the Amazon—including Mato Grosso and the arc of deforestation along the rainforest's southern side—had switched. It was no longer a "sink" for the gas, absorbing more from the air than it released; it had become a "source." This part of the Amazon at least was no longer damping down global warming. The growing pace of savannization and the loss of dense forest meant it was speeding things up. "Each year it gets worse," she said. "It is a self-reinforcing cycle now. We have to stop deforestation while we work out what to do."

Other academics are reaching similar conclusions about the presence of tipping points within tropical rainforests. The forests are resilient, they say, but only up to a point. The more they learn, the more worried they become. A 2021 study found that in the southern Amazon region, including Mato Grosso, areas cleared of forests for big farms like Tanguro had seen temperatures increase three times faster than in areas occupied by small-scale rural settlements.[7]

And this is not just about the Amazon. I read a paper by no fewer than 225 researchers from across the world, who assessed data from more than half a million trees of more than ten thousand species in more than eight hundred forests in twenty-four tropical countries. They found that as soon as mean daytime temperatures in the warmest part of the year hit thirty-two degrees Celsius (90°F), tree growth rapidly declines.[8] Some die; some burn. For every degree above the threshold, carbon emissions from their forests quadruple.

On the southern edge of the Amazon, the mean temperature is beyond that threshold. At Tanguro farm, Coe had said, the mercury regularly rose above thirty-four degrees (93°F). What began as Nobre's modeling prediction for the Amazon is now accepted wisdom for tropical forests as a whole. If pushed too far, we can expect them to die.

⁂

BUT PERHAPS WE CAN AVOID this fate. We should take nothing for granted in the Amazon. For much of my career as a journalist, its destruction had been a regular story and a seemingly unstoppable process. My bulging file of clippings about the Amazon begins with "The Rape of the Amazon" in the *Observer* newspaper in 1979. Through Earth Summits and climate conferences, rainforest declarations and aid programs, the carnage had continued until 2003.

That was the year of the election as Brazilian president of reformist former trade union official Luiz Lula da Silva. Lula may not have been an environmentalist, but he appointed as his environment minister a green heroine, Marina Silva.

An illiterate maid born to a family of rubber tappers in the far west of the Amazon, she had in the mid-1980s met and been radicalized by an internationally famous forest-rights activist, Chico Mendes. When he was assassinated, she bravely took on his mantle, and during her five years in office in Brasília, she had transformed the politics of the Amazon. She brought law and order to a region that had previously been run by land mafias; policed it with real-time satellite monitoring of forest destruction; encouraged states to establish registers of land ownership; and introduced other reforms, such as curbs on the sale of beef and soy from recently deforested land.

It was a propitious moment. After the orgy of land grabs and forest destruction, many of the agribusiness companies who were then converting deforested cattle pastures to soybean farms wanted to secure their investments by staying within the law. They felt that enough land had been cleared. They had made their millions; they craved respectability. Many of them backed Silva's reforms. Brazil combined a growing economy with ever better protection of the rainforest. In the decade after she took office, the rate of deforestation in the Amazon decreased by more than three-quarters. Scientists and environmentalists alike believed that the bad old days were over and that a new era of pragmatic land management was replacing the "Wild West."

Optimism can be a dangerous thing. In 2019, the bad guys hadn't gone away. Their modus operandi remained in place. "They cut and burn the trees. Then they put in cattle, and wait for an amnesty to legalize their annexation, after which they can sell," said Paulo Moutinho, an ecologist who works both at IPAM and with Coe at Woodwell. While such activity was strongly down from two decades before, it hadn't entirely stopped. Around a fifth of the country's soy and beef exports continued to be grown on farms that had carried out deforestation since 2008, said Raoni Rajão, a social scientist I met at the Federal University of Minas Gerais. There were still, he said, plenty of "rotten apples in Brazil's agribusiness."[9] Moreover, those rotten apples had successfully worked to elect a new president who would rescind Silva's Amazon laws.

On taking power in early 2019, Jair Bolsonaro began dismantling the agencies charged with policing the rotten apples, eliminating inspections and telling law enforcement officers to stop issuing fines for illegal logging.[10] Environmentalism was a conspiracy by foreigners to contain Brazil's economic growth, he said. He appointed as his environment minister Ricardo Salles, a right-wing lawyer who promised to repeal Silva's laws, shrink protected areas and open up Indigenous territories to miners. As I traveled the country six months later, criminality was once again the norm out in the jungle. At least ninety percent of the deforestation in the Amazon that year was illegal, Nobre told me.

Bolsonaro also fired the man who told the world what was happening. Ricardo Galvão, director of INPE, which analyzes satellite images of the rainforest, had had the temerity to report the rise in deforestation. He was promptly dismissed. But he had brave lieutenants who were far from cowed. When I called INPE shortly after Galvão's departure to ask for a visit, I expected silence. Instead, Cláudio Almeida, the coordinator of Amazon monitoring, could not have been more obliging. He was keen to show me what he did, and how.

Almeida ran two kinds of monitoring: a long-standing annual check on where forests have disappeared, and a real-time monitor of where fires break out, which Silva had set up in 2003 to enable law enforcers to catch illegal forest clearers red-handed. "I think we are the first country in the world to have such a system," Almeida told me. As well as being a window on the rainforest and a vital tool for law enforcement, it had also proved a scientific gold mine.

"We have logged twelve hundred scientific articles in more than four hundred journals, based on our data," he said. "It is all open to public access. You can find it online. It has been the basis for all public policymaking, and our declarations and promises to climate conferences like in Paris." Obviously, Bolsonaro did not like it.

We watched as one of his analysts sat at a computer screen, comparing current images with those from the previous year, color-coding changes. It was routine work that had recently revealed that almost 2.5 million acres of forest had been lost in the year to August 2019—the highest for a decade and an area roughly the size of Lebanon. His screen watchers also spotted more than 180,000 fires. One night they logged a heavy concentration along the rainforest highway known as BR-163—confirming what farmers had advertised in advance would be the "day of fire," the single most dramatic sign that a new fight for control of Brazil's land was underway.[11] The aim was to grab the hundred million acres of state forests across the Amazon—an area larger than Germany that is neither privately owned nor state-protected. Land speculators see it as ripe for the taking.[12]

Bolsonaro and his backers frame the fight for the Amazon as a battle between the environment and economic development. This is nonsense. The land grabs have little to do with furthering the country's economic development. Having cleared the forest at great ecological cost, the land speculators often barely trouble to make economic use of it. Typical rates of cattle stocking are below one animal for every 2.5 acres.

As Nobre told me over our Sunday lunch, genuine development would involve better use of already deforested land. "If you double up livestock intensity, which is entirely feasible, you could free up more than fifty million hectares [123 million acres] for forest restoration." More productive ranches would also keep some trees. "Cattle don't come from the tropics and they don't like heat. With shade, beef cattle put on more weight and dairy cattle increase milk production," he says. Some ranchers do this, with evident success. Most, however, "still have this Hollywood image of the Wild West. They think they are in Texas. They are not rational. They just want to possess huge amounts of land."

All was not lost, however. Far from it. The land speculators may have regained the political ascendancy, but the big agribusinesses had not gone away. And with their heavily automated farms requiring large investment, they now needed at least a passing relationship with legality, Nobre said. Tanguro farm was a case in point. Its owner, Maggi, had profited from past illegality. He became a billionaire by buying up illegally cleared rainforest. Then, as governor of Mato Grosso two decades ago, the "king of soy" had presided over a surge in deforestation in the state. Times changed, however. Maggi now has all the land he needs and has long been keen to placate his environmental critics. That is why he invited Woodwell researchers to set up a research project on Tanguro farm.

I experienced some of the same agribusiness PR treatment in 2012, when I was for a while courted by the Brazilian Confederation of Agriculture and Livestock. At a personal briefing in London, its president, Kátia Abreu, who later became minister for agriculture, told me with the utmost sincerity that "Brazilian farmers have no interest in clearing land. The main objective today is to produce more with the land already available."

Almost a decade on, the big farmers feared that Bolsonaro's anti-environmental politics could bring about international boycotts of their beef and soy. I mentioned earlier how Bolsonaro grabbed headlines at a UN climate summit in 2019 by raging against European "colonialists" who wanted to tell Brazil what to do with its forests. But Abreu's successor, Marcello Brito, was also in New York that week. He never made the evening news, but he was calling for a halt to the fires and to the presidential inflammatory rhetoric.[13] Soon after, Britain's biggest retailers, including Sainsbury's and Tesco, threatened to boycott Brazilian products if the Brazilian Congress backed a Bolsonaro bill to make it easier for invaders of public forest land to legalize their occupations. Days later, Congress backed off.[14]

Brazilian politics seems as much at a tipping point as the Amazon itself. The two are inextricably linked. We cannot know how things will go. Certainly, 2019 was not a one-off. In 2020, deforestation rates in the Amazon were the highest for twelve years.[15] Amazon fires in June 2021 were the highest for fourteen years. History may see the Bolsonaro years as a nasty blip within a long-term trend towards bringing law and order to the world's largest rainforest, and to halting the runaway repercussions of its loss. Or, on the contrary, it may relate how the dramatic decline in deforestation ushered in by Silva was the aberration, a brief happy interlude in the continued destruction in the Amazon.

— 5 —

# FIRES IN THE FOREST

## *Nature's Way of Starting Over*

AUSTRALIA IS SOMETIMES CALLED "the fire continent." The ecology of the world's driest continent has long been shaped by burning. In the tropical jungles, as we have seen, forests and rains need each other and feed off each other. So in Australia, trees and fire go together. Fire prospers where there are trees to burn. But, strange to say, forests need fires too. Fire prepares the ground and germinates their seeds. In many places, no fires would mean no trees.

That said, even the fire continent had rarely seen anything like the conflagrations that engulfed its southeastern quarter in late 2019 and early 2020. Some said that the synergistic relationship between forests and fires was in danger of being broken. That, rather like the Amazon, Australian forests were approaching a tipping point beyond which the fires would consume the forests for good.

We all saw it on our screens. For four months, flames engulfed large parts of the states of New South Wales and Victoria. More than twelve million acres burned, including a third of the rainforest in New South Wales.[1] At least twenty-eight people died. In Mallacoota, a coastal town in Victoria, a thousand people were rescued from the beach by the Australian navy as the flames closed in. The fires stretched into neighboring South Australia, where the vineyards of the Adelaide Hills, which produce some of the country's most prized and widely exported chardonnays and sauvignon blancs, were reduced to ash. Adding to the mayhem, state authorities shot thousands of feral camels to protect Aboriginal communities besieged by herds coming out of the bush in search of water.

For those concerned about climate change rather than life and limb, the burning trees released approaching 900 million tons of carbon dioxide into the atmosphere—far more than the country's annual emissions from burning fossil fuels.[2] A government report announcing the figures said the forests could be expected to reabsorb an equivalent amount as they regrew. Critics were skeptical. They suggested that would not happen if such fires became more prevalent and the forests would not recover.

Australia is among the countries most exposed to the gathering pace of global warming. Average temperatures rose by two degrees during the twentieth century—twice the global average. The six hottest days ever recorded in Australia, maxing out at 49.9 degrees Celsius (121.8°F), were in 2019. The combination of this unprecedented heat and rainfall forty percent below average resulted in fires more extensive than any of Australia's previous bushfire disasters.

As the country watched the nightly news images in shocked disbelief, the reaction of meteorologists was effectively: we told you so. Thirteen years earlier, the Australian Bureau of Meteorology had warned that, thanks to ongoing climate trends, fire seasons in southeast Australia "will start earlier and end slightly later, while being generally more intense. This effect . . . should be apparent by 2020."[3] Ministers didn't seem to have been listening. In a radio interview as the fires reached their peak, bang on cue, deputy prime minister Michael McCormack dismissed any link to climate change as "the ravings of some pure, enlightened, and woke capital-city greenies."[4] Well, his own scientists, actually.

To be fair, bushfires aren't a new phenomenon. When Captain James Cook reached the continent in 1770, he described it as "this continent of smoke." Its history is littered with tales of the havoc they can bring. The worst incidents have names, among them Black Friday in 1939, Black Tuesday in 1967 and Black Saturday in 2009. Up to a point, they are ecologically beneficial. Much of the continent's ecology depends on regular fires to propagate seeds. Most of its eight-hundred-plus endemic eucalyptus species, which make up around three-quarters of its forests, thrive in fire-prone areas with nutrient-poor soils. Their foliage is rich in oils known as gum (hence their colloquial name: gum trees) that readily burn. The burning releases the trees' seeds from woody capsules and creates areas of nutrient-rich ash where the seeds will germinate.[5] Without fires, there would be no forests.

As the fires raged, some scientists warned of a wildlife apocalypse. Chris Dickman of the University of Sydney hazarded a guess that more than a billion mammals, reptiles and birds could have burned, starved or been eaten by the raptors and feral cats that stalk fire zones. He later estimated that a further two billion had been "displaced." Some endangered species that lived mostly in the fire zones faced extinction, including the eastern bristlebird; the long-footed potoroo, a rabbit-sized marsupial; and the silver-headed antechinus, a mouse-sized carnivorous marsupial that was only discovered in 2013.[6]

Others took a more optimistic view. Just as Australia's forests are adapted to fire, and often need it, so with wildlife. It was wrong to assume that anything living in the fire zone would have died. "Australia's animals have a long and impressive history of co-existing with fire," said Dale Nimmo of Charles Sturt University.[7] Some creatures had evolved well-developed escape routines. Others know to hunker down in deep burrows, or go into temporary hibernation until the fires are gone. Creatures such as raptors may even benefit from the fires, with more prey on the run.

While fire may be a necessary factor in the life cycle of some of Australia's ecology, the scale of the 2019–20 fires was almost unprecedented. In a typical year, fires burn two percent of the forests of southeast Australia, but the 2020 inferno burned over twenty percent.[8] It was so hot and dry that the fires spread into forests with eucalyptus species that needed wetter, fire-free conditions to survive. They spread to places where there is no synergy between fire and forests. How well the landscape as a whole will recover will be a critical question for forest ecologists.[9]

THE AUSTRALIAN FIRES GAINED international attention in part because they reflected what looked like a global pattern. Fires in the Amazon the previous August had also made headlines. There were also extensive fires in 2019 in the boreal forests of Siberia, where some ten million acres burned. Some four million acres went up in flames in the rainforests of Indonesia. In 2017, 2018, 2019 and again in 2020, California and the rest of the American Pacific coast were struck by some of the deadliest and most extensive wildfires ever recorded in the region. They covered millions of acres, caused dozens of fatalities and forced the evacuation of hundreds of thousands of residents in the forested hills.

Further north in Canada it has been the same story. In 2016, wildfires in spruce plantations amid the boreal forests of northern Alberta almost wiped out the oil town of Fort McMurray, forcing the evacuation of eighteen thousand people. The conflagration was reportedly the costliest natural disaster in Canadian history. To the east, there has been a reduction in wildfires in the past two decades. But in British Columbia and further north, deep into the Arctic Circle in the Northwest Territories, fire activity has been unprecedented.

What is going on? Climate change is undoubtedly increasing the risks of fires worldwide.[10] There is more of what meteorologists call fire weather. "Human-induced warming has already led to a global increase in the frequency and severity of fire weather, increasing the risk of wildfires," says Matthew Jones of the UK-based Tyndall Centre for Climate Change Research.[11] David Bowman, a fire ecologist at the University of Tasmania, found an alarming nineteen-percent increase in the average length of drought seasons with fire weather since 1979. We have seen too how Mato Grosso and other parts of Brazil's arc of deforestation could be on the cusp of tipping from rainforest to savannah, driven by a lengthening dry season and more fires. Paulo Brando of the Woodwell Climate Research Center calls it "the gathering firestorm."

Many other fire-prone forest ecosystems could be on the verge of becoming grassland. Over the past half century, the number of wildfires covering more than a thousand acres in the American West has increased fivefold. It is a trend that researchers link to a doubling in the number of fire-weather days.[12] In Mediterranean Europe, from Portugal to Greece, wildfires are an increasing summer threat. Few doubt that fires will also become more frequent in the boreal forests of the far north, as the Arctic warms.

More fire weather does not mean there are more fires, however. The actual extent of fires around the world has been falling. While at the NASA Goddard Space Flight Center, Niels Andela found a twenty-four-percent decline in two decades, largely due to fewer fires being set by farmers to clear bush in the increasingly densely populated savannah grasslands of Africa.[13] What has increased is wildfires—uncontrolled, unanticipated and unwanted fires, ripping through ecosystems that are sometimes appreciative of the blazes, but often not.

Wildfires may be an inconvenience for humans, especially when they are large and untamed. But nature needs them. If humans don't set them, then lightning probably will. "Fire is not an end but a beginning," says Dominick DellaSala, fire guru and chief scientist at the Geos Institute in Ashland, Oregon.[14] "It's nature's phoenix." Even the biggest, hottest fires do not kill all trees, and the heat germinates seeds lurking in the soil. Meanwhile, the burning creates clearances that let in light and encourage new growth. Fire "shakes and bakes; it frees nutrients and restructures biotas," says Stephen Pyne, a fire historian at Arizona State University, who was once himself a firefighter.[15] A 2021 study in Alaska found that, counterintuitively, wildfires could even benefit carbon storage; in a warming world, researchers found that slow-growing spruce trees were often being naturally replaced by faster-growing aspen and birch forests that would eventually store more than five times more carbon in their soils.[16]

Fire is essential to forests. But we have created problems for ourselves by messing with the natural fire cycles. In successfully preventing fires, we have over decades inadvertently stored up prodigious quantities of kindling in forests—ready to supercharge the first fire that gets out of control. Hotter, drier conditions have exacerbated the situation, and the result is a plague of fires.

Managers of parks and other wildlife reserves have in some places gotten wise to this. They set controlled fires at the start of the dry season, which work their way across the ground consuming dead wood. The idea is to prevent big fires at the end of the dry season that will rise up and rush through the canopy. This is a return to the old ways of Native Americans, says Lee Klinger, formerly at the National Center for Atmospheric Research and now an independent fire researcher in California. In North America before Europeans took over, "native people burned extensively. They burned frequently, and so less destructively." They did it to clear brush; to maintain clearings for pastures and croplands; to stop conifers taking over from mixed woodlands; to entice game; to encourage bushes for berries and grasses for basket-weaving, both of which require open burned areas; and to limit the spread of wildfires. It worked.

The Karuk, Miwok, Hupa and Yurok in northern California all still know what "good fires" are, and that they prevent "bad fires." But for most of the past century, federal and state authorities largely outlawed them. Their cardinal rule, adopted from European practice, was to fight and prevent all fires. Even environmentalists, in a long tradition stretching back to Aldo Leopold, agreed. But they were wrong. Outright fire suppression is ecologically damaging and ultimately self-defeating.

Most forests in the West require fires—not the superhot infernos rushing through canopies they have to contend with now, but the regular annual repeat of cooler ground fires that return nutrients to the soil, aid seed germination and create space and light for the trees that need it. "While fires can be dangerous, they are inevitable and necessary in many ecosystems, and humans have long adapted to them," says Sara Worl of the University of Oregon.[17]

It took European settlers a long time to get wise to the virtues of fire, she says. Centuries, in fact. They were too busy colonizing and conquering to consider how the locals successfully managed their land. "They were struck by the rich biodiversity of California's forests, woodlands and prairies, but they didn't understand that Indigenous peoples' use of fire was responsible for them. Instead they sought to suppress fires wherever possible." But, as Pyne puts it: "It turns out that the educated elite were wrong and the nominal 'primitives' right."

Today the doctrinal divide between Indigenous and non-Indigenous approaches is breaking down, with Indigenous wisdom prevailing. The official rethink began in the 1980s, when forest ecologists in California noticed how, in the absence of fires in the forests, the state's giant sequoia trees were not reproducing. Their cones required the heat of fire to open up and release seeds that germinated in freshly burned soil. So no fires meant no more trees.

The case for burning is now much better understood. Grassroots associations of locals in parts of the West have organized themselves to carry out controlled, or "prescribed," burns on private land. The groups are disparate. In 2019, a *Mother Jones* magazine reporter noted that one event she attended saw "Trump-supporting ranchers" setting fires with "retired hippies and back-to-the-landers, logging workers and Save the Redwoods League activists."[18] Typically, the fire-setters are guided by Indigenous fire experts, such as Bill Tripp of the Karuk Tribe Department of Natural Resources. Tripp is also a member of the Indigenous Peoples Burning Network, an alliance established with the assistance of The Nature Conservancy.[19] Fire scientist Crystal Kolden of the University of Idaho says the network is "restoring ecological function to fire-adapted ecosystems ... following a century of fire exclusion."[20]

So far as Native Americans are concerned, the return of controlled burning is at least as much a cultural as an ecological restoration. The Karuk see fire as a medicine for the forest, says Frank Lake, an ecologist for the US Forest Service who is of Karuk descent.[21] Indigenous communities often talk about "cultural burning" as a profound connection between fire and their lives, says Margo Robbins of the Yurok reservation's Cultural Fire Management Council. "Fire has the ability to re-establish that connection." Karuk fire management systems maintain diverse forests that provide them with fruit from the forest and grasslands where their elk can graze.

The sharing of fire management skill is spreading far beyond California and the American West. The Indigenous Peoples Burning Network has swapped expertise with Aboriginal peoples in Australia, including an exchange visit with the Martu bush burners in the deserts of Western Australia.

But while restoring the link between fires and culture comes naturally to many Native Americans, it is much less easy for those brought up with European traditions of fire suppression. Despite its evident ecological benefits—and despite growing anger about the destructiveness of wildfires caused by the accumulation of fuel—gaining public and regulatory acceptance of the need for an accommodation with fire is proving painfully slow. Some foresters still believe in suppressing all fires. Many citizens, especially people who have set up home in the forested mountains, fear both fire "escapes" during controlled burns, and the health consequences of smoke from forest fires in their neighborhoods. As a result, "the best available science is not being adopted," says Kolden.[22] The overall amount of prescribed burning has barely increased in the West since the start of the twenty-first century. Kolden figures California should be doing five times more than it does.

One barrier to the new thinking is an understandable perception that climate change is the real culprit in the spread of wildfires. As his state burned in late 2020, the governor of Washington, Jay Inslee, declared: "This is not an act of God. This has happened because we have changed the climate."[23]

It certainly plays a part in these newly intense conflagrations.[24] As we have seen, there are more "fire days" in the West today, with higher temperatures and more droughts. But to blame climate change alone is to ignore the fact that fire suppression over many decades has hugely increased the amount of fuel on the forest floor, making forests dramatically more vulnerable to fire. Of course we should halt climate change. No question. But without a radical increase in controlled fires, "more catastrophic wildfires are inevitable," says Kolden.

The question now, in places where fire risks are increasing, is whether a rekindling of more traditional methods of fire management can hold back the kinds of dangerous wildfires that neither humans nor the natural environment is equipped to deal with. The kinds of fires that exceed the resilience of ecosystems and tip them into new states, whether it be savannization in the Amazon or desert formation in Australia. "Forest and savannah are alternative stable states for tropical land," Partha Dasgupta of the University of Cambridge reported in a 2020 review of the economics of biodiversity for the British government. Once high temperatures and lack of rain trigger fires, there is a runaway effect, "making the return shift to forest extremely unlikely."[25]

Australia has some history here. What we saw in the fires of 2019–20 may actually have been a continuation of a process of deforestation and fire triggering climate change that started long ago. When humans first arrived on the continent, it was a lot wetter and greener. One theory is that it was humans who turned it into the world's driest continent, through the way they managed the bush.

Gifford Miller of the University of Colorado has pieced together a chronology of events there around 45,000 years ago. Back then, the arid Nullarbor Plain in South Australia, whose name is Latin for "no trees," was home to forests of eucalyptus trees clambered over by tree-dwelling kangaroos. Inland, across what is now known as the outback, today's desert depressions were filled with huge lakes. Lake Eyre, also known as Kati Thanda, drains an area the size of France and Germany combined. Today it is normally a dry, salt-encrusted plain, but once it was a permanent body of deep water that extended for around 2.5 million acres.

What happened? Meteorologically, it seems that the annual monsoon winds that brought moisture from the ocean into the interior dried out. Global climate factors cannot explain this, says Miller.[26] "The only variable that changed is humans colonized the continent." He argues that the newcomers hunted by burning bush to round up their prey and that this systematic burning pushed the wet, forested ecosystem beyond its limits. By reducing vegetation, they reduced the transpiration that kept the winds wet. The monsoon winds brought less moisture into the continental interior, which in turn caused more loss of vegetation and even less rainfall. The old ecosystem suffered a cascading collapse. So today, rather than having a wet interior, rainfall "diminishes rapidly inland," to less than twelve inches within a few hundred miles of the coast, notes Miller. The modern outback is arid, and it was humans who caused it.

The stories of recent events in Australia and the Amazon certainly offer a warning about the dangers of passing tipping points in forests. But they are the flipside of the bigger story of how amazingly resilient and adaptable forests are. Through transpiration, flying rivers and their chemical breath, forests have the power to manage their own environments for their own benefit. Forests long ago made our planet's atmosphere, environment and life-support systems. And they still do it. We mess with their life-support systems at our peril.

PART II

# FROM PARADISE TO PLUNDER

---

There are no pristine forests. Wherever we look—deep in the Amazon, in the heart of the Congo basin or up the Orinoco—we find human footprints. Great ancient civilizations flourished amid the trees. There was clearing, but almost always the jungles regrew. We successfully cohabited with trees, harvesting them without destroying the forests. But in recent times nature's resilience has struggled to keep up with our seemingly unquenchable thirst for rubber and mahogany, soy and palm oil, paper and beef. From the thorn forests of Paraguay to the peat swamps of Borneo, we have downed millions of trees to meet this demand. Part Two is a roller-coaster ride through history, before we take stock of the damage, find where and how forests persist, and ask if nature can return.

---

# —6—

# LOST WORLDS

## *Pre-Columbian Cities That Gardened the Rainforests*

THE FIRST EUROPEANS to enter the Amazon rainforest sent back accounts of cities, roads and fertile fields along the banks of its major rivers. "There was one town that stretched for fifteen miles without any space from house to house, which was a marvelous thing to behold," wrote Gaspar de Carvajal, a Dominican missionary and chronicler who traveled with Spanish conquistador Francisco de Orellana on his journey up the Amazon in 1542.[1] Beyond the city limits, there were orchards and fields. "The land is as fertile and as normal in appearance as our Spain," he reported.

Such accounts were later dismissed as inventions, not least because no later expeditions reported seeing such things, but recent archaeological research has shown that the chroniclers were not fantasists. It is our romantic notions of a pristine rainforest wilderness that are adrift from reality. In the fifteenth century, there were indeed substantial urban settlements on the banks of the great rivers of the Amazon rainforest. Their inhabitants were not hunters and gatherers, but sophisticated farmers who managed their soils, had permanent fields and turned many areas of forest into what amounted to orchards.

Much of what is today the world's largest tract of apparently untouched rainforest was, until little more than five hundred years ago, a landscape dominated by human activity, says Charles Clement of Brazil's National Institute of Amazonian Research in Manaus. For more than ten thousand years, its inhabitants had been remaking the forest, planting what they wanted, culling what they did not, burning, farming, draining, ponding and fertilizing the soil, and creating thousands of villages and dozens of urban settlements. "Few, if any, pristine landscapes remained in 1492," when European colonists arrived, says Clement. "Many present Amazon forests, while seemingly natural, are domesticated."[2]

The notion of a pristine Amazon is a legacy of chronicles written in the eighteenth and nineteenth centuries by explorers who were describing what was left more than two centuries after the arrival of the first Europeans. That initial invasion unleashed a human holocaust in which more than ninety percent of the Amazon population was wiped out by diseases and superior weaponry. The survivors had retreated into the bush, leaving the jungle to reclaim their fields, orchards, palaces and plazas.

The evidence for this radical rethink of the Amazon has been stacking up for some time. Thanks to a combination of deforestation and remote sensing, archaeologists have uncovered dense urban centers that would have been home to up to ten thousand inhabitants. One of the first to stumble on the story was Anna Roosevelt of the University of Illinois at Chicago. She was the great-granddaughter of President Theodore Roosevelt, who had himself gone to the Amazon in 1913, after retiring from office, in search of "the last great wilderness on Earth." Seventy years later, she went digging beneath the trees of Marajó, a large flat island at the mouth of the Amazon, and found massive earth mounds connected by causeways and canals. She had unearthed part of a pre-Columbian "lost civilization."[3]

A decade later, Michael Heckenberger of the University of Florida found similar mounds hundreds of miles to the south in the

Upper Xingu valley, quite near Tanguro farm. There were nineteen settlements, with castle-like structures, bridges and dikes, orchards, manioc gardens and open parklands, all linked by straight roads up to 150 feet wide, suggesting a strong central authority, he said.[4] The "Lost City of Z" may not be such a myth after all.

Modern analysis of satellite and radar images has often yielded the first signs of such startling rainforest remains. Sometimes, however, discoveries are made by happenstance. Back in the 1950s, a Texan engineer searching for oil drove a beach buggy across the Llanos de Mojos in the lowlands of northern Bolivia. The ride was unexpectedly bumpy and the bumps seemed to be at regular intervals. The entire landscape was corrugated. Kenneth Lee formed a theory that he had stumbled on the remains of some vast agricultural civilization. It sounded eccentric at the time, but was proved to be true.

The former forest lands turned out to hide a network of ridges and canals covering hundreds of thousands of acres, whose construction would have required engineering "comparable in scale to the pyramids," according to the first researcher to map them, Clark Erickson of the University of Pennsylvania. The raised ridges kept crops like manioc, maize and squash clear of floods and frosts, while the canals beside them delivered water for irrigation in the dry season. This corrugated landscape was punctuated by raised earth causeways connecting circular forest islands, where people lived surrounded by trees.[5]

People "completely altered a landscape that is today empty and abandoned, save for regrowth of forest," Erickson told me after making his discoveries in the early 1990s. An archaeological study published in 2020 confirmed that manioc and squash were being cultivated here ten thousand years ago, around more than four thousand of these artificial raised forest islands.[6] This made the Amazon one of the earliest-known locations of crop domestication anywhere in the world. Similar landscapes have since been found amid forests on the Guiana coast, in northeast Brazil, in the Xingu basin and along the Orinoco River in Venezuela.

Lee, their first discoverer, became so enthused by what he'd found that he changed career, moved to Bolivia and spent the rest of his life studying local archaeology and ancient culture. After his death in 1999, locals set up an archaeology museum in his name. It was a fitting tribute, but the heritage of such lost civilizations was never entirely lost, and it exists outside museums. In fact, if you know how to look, it is all around the Amazon and beyond.

Today's villages are often founded on the sites of old, much larger settlements, and today's forest trails still follow paths through ancient modified forest-scapes, containing Brazil nuts, acai palms and other valuable tree species planted and domesticated for thousands of years. Erickson has cataloged more than eighty plants known to have been bred and cultivated, including cacao, sweet potatoes, pineapples, peppers, peach palms and tobacco. Amazon forests "were more similar to gardens, orchards and game preserves than wilderness," he says. "Most landscapes that are today appreciated for their high biodiversity have evidence of human use." The diversity is "a result of, rather than in spite of, long human disturbance."[7]

The hand of human cultivators seems the most likely explanation for one of the conundrums about the distribution of tree species in the Amazon. Ecologists have documented sixteen thousand tree species there. Among them, however, just 227 "hyperdominant" species account for half of the total, even though there is no evidence that they are in any way better suited ecologically to their environment than the also-rans.[8] What distinguishes most of them is their value to humans.

These findings have been known about since the 1990s, yet they don't seem to have dislodged the common conception of the Amazon rainforest as a pristine environment. "Reports of empty forests continue to captivate scientific and popular media," says Clement.[9]

We cling to our myths of an unsullied Eden. Almost any rainforest is routinely described as pristine. But these are myths with consequences, says Clement. To deny the human story of the Amazon is, he says, to perpetuate the "notion that the native Amazonian people are somehow noble savages" and to deny the descendants of those past societies both their history and their right to manage their ancestral home.

In a famous paper called "The Pristine Myth," published on the five hundredth anniversary of Columbus's arrival in the Americas, William Denevan of the University of Wisconsin–Madison called the Amazon a "humanized landscape almost everywhere... There are no virgin tropical forests today."[10] But his canvas was not just South America. It was North America too. Wherever he looked, from the Canadian Arctic to the Deep South, and from New England to California, he found evidence not just of human occupation, but of transformation of the land. That included large-scale management of forests, primarily through the use of fire that we discussed in the previous chapter.

Once considered controversial, this revisionist view of American landscapes is now widely accepted among researchers, if not always among environmentalists or laypeople. Archaeologists have found dotted across the Midwest, from the Great Lakes to the Gulf of Mexico, urban settlements occupying thousands of mounds sometimes tens of feet high and covering several acres each. Many of these former centers of population are surrounded by land that was extensively engineered for drainage or irrigation—a product of what James Parsons of the University of California, Berkeley, has called "a mania for earth moving on a grand scale [that] runs through much of New World prehistory."[11]

The pre-Columbian population of North America may have numbered as many as a hundred million. These industrious folk often turned forests into mosaics of meadows and open woodland that was conducive to fire-tolerant and light-loving trees and to berry bushes that attracted wildlife. In the clearings, they grew corn. The forests were so open and parklike that you could drive a coach and horses through them, according to early colonists from Europe.[12] Even the prairies, which are regarded today as quintessential wild American landscapes, may be largely anthropogenic creations, maintained by fire in places where the rainfall would otherwise have supported forests. In effect, they were "vast buffalo farms," says author Charles Mann.[13]

But, as in South America, the decimation of native populations following the arrival of Europeans ended this landscape stewardship. Their diseases and destruction resulted in a dramatic return of forests in many places. A new wilderness grew that later European settlers mistook for ancient virgin forest. Many of modern America's problems in managing its forests and parks—in particular dealing with fire—have arisen from that misconception.

Denevan was talking about the Americas, North and South. But the rethink about American landscapes is mirrored in many other regions of the world, including the rainforests of Central America, the Congo basin and Southeast Asia. Everywhere, it now seems, much that appears as pristine and primeval jungle is actually more likely long-abandoned and overgrown gardens.

THE MAYA PYRAMIDS of Guatemala are now tourist attractions. The civilization that built them lasted for more than two thousand years and was a center of science and engineering, religion and astronomy, hydrology and agriculture. It abruptly disappeared twelve hundred years ago. Overpopulation, changing climate and deforestation have all been blamed. The current theory is that the Maya exacerbated natural drought by replacing forests with crops, which transpired less. But the Maya may not have deforested much at all. "They appear to have gardened the local forest, rather than practicing forest clearance," says Patrick Roberts of the Max Planck

Institute for the Science of Human History in Jena, Germany.[14] Certainly, if the forests ever went away, they came back within a century or so, and today still contain an abundance of tree species that were widely cultivated by the Maya for timber, fruit and, as we shall see later, even chewing gum.

Likewise in Southeast Asia, the temples of Angkor are the remains of what at its height in the twelfth century was probably the largest pre-industrial urban complex in the world, stretching across a quarter-million acres of former forest.[15] The forest has done its best to return. Tourists visiting Cambodia to marvel at the great temples often take selfies next to giant tree buttresses pushing over walls and gouging out foundations. But the forests of Southeast Asia were transformed at least as much by agricultural societies. Many of the trees on the forested islands of modern-day Indonesia and the Philippines have been cultivated and are often non-native species. As early as ten thousand years ago, farmers had shipped sago palm from New Guinea to Borneo, where it remains a vital subsistence crop to this day. "The forests of this vast region are, to an extent, a cultural artefact," says Chris Hunt of Queen's University Belfast.[16]

From New Guinea to Thailand and the Philippines to Java, the new narrative is being borne out. Most of Borneo's "pristine" forests are secondary growth, returning to cover regions once extensively cultivated with sago palm and other domesticated crops. "Much of the low altitude to lower montane vegetation of [both the mainland and islands of Southeast Asia is] a long-established cultural artefact," says Hunt.

Africa has its equivalents of Angkor Wat and the Maya pyramids, though they are much less celebrated. Forgotten, even. A few years back, I went in search of earth ramparts that once surrounded the kingdom of Ijebu, which flourished amid the forests of southern Nigeria. The earthworks extend for a hundred miles and enclose an area of land the size of Greater London. They were in places seven stories high, complete with guardhouses, moats and garrison barracks. Known as Sungbo's Eredo, they were at the time of my visit thought to have been dug early in the ninth century—around the time that Charlemagne ruled in Europe—by Oloye Bilikisu Sungbo, a fabulously wealthy Ijebu noblewoman. There was once a sacred grove that contained a shrine dedicated to the Queen of Sheba here. More recent archaeological digs suggest that the earthworks may date back five thousand years.[17] Whatever their age, they shatter the old myth that Africans were primitive hunters, gatherers and shifting cultivators until Europeans showed up.

I was tipped off about Sungbo's Eredo by Patrick Darling, a geographer at Bournemouth University, who died in 2016.[18] He first stumbled on the remains of these great earthworks while driving in the area. Until Darling's visit, almost no one outside the region knew about them, and he began a campaign to have them protected. What most impressed him, he told me, was the skill displayed by the Ijebu in surveying and construction. "Without compass or aerial photographs, they were able to keep the ramparts on course whatever the obstacle," he said. "They had the ability to retain a coherent masterplan in thick rainforest, frequently interrupted by great fingers of swamp."

Despite Darling's efforts, these grand earthworks remain mostly unknown. Though a few pilgrims visit Sungbo's shrine at Christmas, many of the ramparts are inaccessible. I spent an afternoon looking for them, after visiting an agricultural research station in nearby Ibadan. But the scientist who had promised to guide me seemed spooked by the enterprise and failed to show. The ramparts that I found were piles of earth covered in green moss and stretching into boggy bush. The ditches reeked of rotting vegetation and echoed with the sound of croaking frogs. Nobody was taking care of them and no one else was visiting. That is a shame. They probably have a story to tell about living and prospering in the rainforest that is at least as grand as that of the Maya pyramids or the Angkor temples.

Much of Africa's ancient history is similarly lost, or restricted to academic studies. The Congo rainforest of Central Africa, the world's second largest, appears pristine to modern eyes. That is deceptive. The forest floor contains the remains of extensive agricultural and urban activity, and even of industrial societies. Pollen in swamp sediments provides evidence of widespread deforestation between three thousand and two thousand years ago.[19] Some researchers blame drought for the disappearing trees. That may have played a part, but it appears that long before the Romans were in charge of North Africa, Bantu migrating into the Congo basin were chopping down trees to fuel iron smelters, and even planting oil palms in the clearings.

Michael Fay of the Wildlife Conservation Society, who has traveled the area extensively on foot, says that soils across one of the least populated regions on the planet are impregnated by oil palm nuts, radiocarbon-dated to the era of deforestation. "Extermination of the forest likely occurred over vast areas," says French archaeologist Richard Oslisly, of the Institute of Research for Development in Marseilles.[20] However, Fay says people may have successfully operated their activities within forests without destroying them. Nobody knows, nor is it clear, why the population subsequently crashed. Recent evidence suggests a major epidemic of some kind among Bantu-speaking people in the region around fifteen hundred years ago. At any rate, according to Fay, it does seem that "*Homo sapiens* is probably responsible for much of the vegetation currently seen in Central Africa, even in seemingly virgin areas."

The imprints of those past uses of forests remain in the landscape, if you know how to look. Increasing numbers of the superficially natural assemblages of tree species turn out to be the leftovers of past cultivations. Perhaps even more remarkably, so are many forest soils. Lots of tropical forest soils, particularly around former settlements, have been mulched and composted. They contain all kinds of organic detritus of village life: plant waste, charcoal, feces, fish scales and shells, animal bones and even turtle backs, making them richer in nutrients and soil microorganisms—a stark contrast to the generally thin, infertile soils of most rainforests.[21]

First discovered by outside scientists in Brazil, these human-made soils are named *terra preta*, Portuguese for "dark earth." They may cover as much as fifteen million acres of the Amazon, in thousands of individual patches typically only a few tens of acres in size, though often more than three feet thick. Their charcoal is a key ingredient. It can stick around for more than a millennium and its porous structure traps nutrients. "You might expect this precious fertile resource to be found in the deep jungle, far from human settlements or farmers," says James Fraser, an anthropologist at Lancaster University. "But I go looking around the edge of villages and ancient towns, and in traditionally farmed areas. It's usually there. And the older and larger the settlement, the more dark earth there is."

Most *terra preta* in the Amazon dates from pre-Columbian times. Its value was probably first discovered by chance, when people realized that crops grew best amid organic waste. There seems little doubt, though, that subsequently it was often deliberately created, and constantly added to. Anthropologists say that the superior fertility of dark earth explains how dense urban populations in the Amazon were able to feed themselves.

The practice was probably largely abandoned when populations collapsed after European colonization. But in some places it would have continued in use, and knowledge about it never entirely disappeared. In modern times, many communities have returned to *terra preta*. Clement studied this in detail along the Madeira River, the largest tributary of the Amazon. Most farmers tilled the dark earth found largely along the riverbanks, he reported. The farmers told him that it allowed for much shorter fallow periods.[22]

Dark earth was once thought of as a unique Amazon technology. But it seems that rural communities in other tropical rainforests with poor soils developed the same strategy for improving their soils with organic waste. After several years working in the Amazon, James Fraser moved to Liberia to look for dark earth in the last substantially forested country in West Africa. I met him in the capital, Monrovia, during a short break after he had set up house in a remote village in Lofa, the country's northernmost county. He told me that within weeks of arriving, he had been taken to 150 patches of dark soil clustered around Wenwuta village and its neighbors. These soils were much more fertile than the natural iron-red laterites. He judged by the amount of pottery shards in them that some patches must be hundreds of years old.[23]

"The villagers know all about dark soil," he told me. "It is understood as an inevitable and fortuitous consequence of settled life—and one that produces a valuable resource." The village women were still actively cultivating more, by collecting their food waste and putting it on soils to create new kitchen gardens. "If you look closely in the records of African explorers and agronomists in the nineteenth and twentieth centuries, you can find occasional reports of such soils," said Fraser. Their diaries reported that in places farmers burned wood and other vegetation under a covering of soil and then distributed the resulting ash across their fields. Until recently, no outsiders looked for a pattern in these reports or wondered about their significance. The best soils in Liberia, it turns out, are human-made. One villager had told him, "God made the soil, but we made it fertile."[24]

Once researchers started looking for dark earth in forests, it turned up everywhere. Fraser found examples over the border from Liberia in Sierra Leone. Kojo Amanor, a land-use researcher at the University of Ghana, mapped it around abandoned villages in central Ghana. Similar findings have been reported in Guinea, Chad, Cameroon, Malawi, Congo-Brazzaville and Ethiopia.[25] They may be in Southeast Asia too. Douglas Sheil of the Norwegian University of Life Sciences saw examples near villages in Indonesian

Borneo. "Similarities include riverside locations, soil characteristics such as high phosphorus levels and improved fertility in comparison to neighboring soils." Local people, while valuing them for growing crops, "were unaware of their origins." But, he concluded, human activity "appears to be the only plausible explanation" for their presence.[26]

*Homo sapiens* has been actively manipulating tropical forest ecologies for many thousands of years. Erle Ellis of the University of Maryland estimates that even six thousand years ago, nearly half of the ice-free land surface of the Earth was in some form of use for agriculture. Shifting cultivation was probably the norm in forests.[27] "Extensive settlement networks clearly persisted in the tropical forests of Amazonia, Southeast Asia and Mesoamerica, for much longer than modern industrial and urban settlements in these environments," says Patrick Roberts. Our ancestors didn't just live in the rainforests, they reorganized them "in fundamental ways, with outcomes that have affected the natural histories of these forests to the present day."[28]

This new history of our relationship with forests should change how we think about them today. If most apparently pristine forests are in fact regrowth from past human activity, then the conventional idea of deforestation as a one-way street with no going back does not hold. It shows that forests are not the fragile ecosystems we imagine, but resilient and often fully able to recover from disturbance. But they have to be given room. So the tragedy of the modern-day plunder of our forests is not just the scale of its loss, but the fact that the land is being taken over for other purposes that preclude recovery. Before we move on to consider how to achieve the great restoration, the next few chapters investigate what we have done to our forests, in a plunder of nature without precedent.

— 7 —

# THE WOODCHOPPER'S BALL

## *Post-Columbian Pillage and Roads to Ruin*

WHEN HERNÁN CORTÉS VISITED the Aztec court of Montezuma in Mexico, he noted before destroying the place how his hosts prized a product of the local cacao bean, the raw material for chocolate. The king "took no other beverage than the *chocolatl*, a potation of chocolate, flavored with vanilla and spices, and so prepared as to be reduced to a froth of the consistency of honey, which gradually dissolved in the mouth and was taken cold."[1] Cortés's interest in this forest bean was no doubt further aroused by the fact that Montezuma drank his *chocolatl* before entering his harem, suggesting that it was an aphrodisiac. The Aztecs mixed their drugs promiscuously. It was claimed that at Montezuma's coronation in 1502, two decades before Cortés spoilt the party, the new king gave his guests *chocolatl* mixed with the local psychedelic mushroom, *Psilocybe*.

A few decades later, the Spanish physician Francisco Hernández reported to King Philip II on how the Aztecs used other native forest plants for medicines and flavorings. They cultivated many of them in the botanical gardens of the capital city, Tenochtitlan. Some of these plants eventually made it to Europe. Queen Elizabeth of England became partial to vanilla. Meanwhile, conquistadors in Peru stumbled on the stimulant properties of coca leaves. They fed them to slaves to keep up their work rate in silver mines. Coca only caught on in Europe in the nineteenth century, after a German chemist isolated an alkaloid from their leaves. After which, cocaine became all the rage, celebrated by everyone from Émile Zola to Buffalo Bill, Queen Victoria to Jules Verne, and Arthur Conan Doyle to the cardinals of the Vatican.

There were poisons to be had in the jungles. French scientist-adventurer Charles-Marie de La Condamine stumbled on the use of curare among Amazon Indigenous people. It was a mixture of poisons extracted from the bark of several trees, sometimes with added snake venom and essence of ants, then boiled for two days and strained to make a deadly paste for tipping arrows or putting in blowguns. Its great advantage was that while it killed hunted animals, it did not taint the meat.

The forests also yielded medicines. At some time early in the seventeenth century, Jesuit missionaries in the Peruvian Amazon adopted a native forest treatment for fever, made from the bark of a tree found among the orchids in the cloud forests of the Andes. According to legend, Lady Ana, the Countess of Chinchón and wife of the local viceroy, became the first European to be cured of malaria using an infusion of the bark, which contained a bitter-tasting alkaloid named quinine. The story of the countess's cure may be a myth, but it was sealed when taxonomist Carl Linnaeus called the tree cinchona after her. For two hundred years, it was the only known cure for malaria, the scourge of the tropics.[2] Without it, much of the story of imperial invasion of the tropics and its rainforests would be different. And it was only available as a result of cutting down trees in the Andes.

Cinchona notwithstanding, rubber was the greatest industrial product extracted from the jungles of South America by Europeans. The first news of a strange milky gunk that oozed from cut bark reached Europe from Christopher Columbus, who said he saw children playing with balls made from the gum of a tree on Hispaniola. Others found that the Quechua people in Peru used the latex to waterproof cloth and make ropes and bottles. It would be another two centuries, however, before La Condamine told British chemist Joseph Priestley about it. Priestley, the discoverer of oxygen, played around in his lab and noticed that, when hard, the substance would rub out pencil marks. He called it, with admirable directness, rubber.

The name stuck. Soon there were rubber boots and bands. Charles Macintosh improved its waterproofing qualities by mixing it with naphtha, a waste gas given off when distilling tar. He made a light waterproof cloth that was soon used in kit bags, air beds and the mackintosh raincoat. Then a bankrupt American inventor called Charles Goodyear mixed it with sulfur to render it flexible enough to make rubber tires, followed by fire hoses, catheters and condoms.

Soon tens of thousands of tons of latex were being exported annually from the Amazon ports of Manaus and Belém. For a while, thanks to the rubber boom, the Amazon was the destination of choice for adventurers. In 1855, before he ever made his name as the chronicler of Tom Sawyer and Huck Finn, Mark Twain stood on the docks of New Orleans asking about ships headed for Belém. "I was fired with a longing to ascend the Amazon and to open up a trade with all the world," he wrote in an essay, "The Turning-Point of My Life."[3] Manaus, the metropolis at the heart of the Amazon, was known as the Paris of the tropics. It boasted an opera house copied from La Scala in Milan and supported eight daily newspapers. The story goes that its high society sent their dirty laundry to Lisbon and Paris.

Such riches came at a terrible cost. Forests in Acre, a region in the far west of the Amazon basin, contained the most rubber trees. Those forests became for a while among the most valuable slices of rural real estate in the world, with tens of thousands of migrant rubber tappers. The area was already disputed between Bolivia and Brazil even before the rubber boom, and the two countries fought a war over the territory at the turn of the century, which Brazil won. Despotic colonies formed in these wild outposts. Reputedly, the worst was run by a Peruvian entrepreneur called Julio César Arana on the Putumayo River, a major Amazon tributary. Here, in a fiefdom area the size of Belgium, he employed native people in chain gangs. If they failed to deliver their quotas of latex, they were strung up or burned alive. Women were kept in breeding farms to supply their replacements.

The suffering in Acre was more than matched by what European colonialists were getting up to in the jungles of Central Africa, where the "Belgian" Congo was run by the Belgian king Leopold II from 1885 to 1908 as a personal estate. He sent in a private mercenary army to strip it of its elephant ivory, around half of which was shipped to Britain to be carved into knife handles, combs, fans, brooches, chess pieces, snuff boxes, piano keys, billiard balls and much else.[4] Soon, Africans were rounded up to do the killing. Hundreds of thousands of them. Elephant numbers began to dwindle. When the slaves failed to deliver enough ivory, they were lashed with a chicotte—a whip made of sun-dried hippopotamus hide cut into long sharp-edged strips. A hundred lashes were usually fatal.

Leopold's lust was not sated even then. After seeing the Amazon rubber boom in full spate, he sought African vines with a latex that could be made into rubber. He found them. But while Amazon rubber tappers took the sap without killing the tree, Leopold's workers were instructed to rip up the African vines, because that way they delivered more latex. This wrecked vast swathes of the forest. As harvests diminished, whole villages were taken hostage, women raped and men killed. Some estimates put the number killed during Leopold's "rubber terror" as high as eight million. The local British consul, Roger Casement—who also played a part in exposing the savagery of the rubber camps in Brazil—reported back to London on the depopulation of whole areas of the country. It only ended when the Belgian government annexed the land from their king in 1908.

Colonial plunder of the riches of the tropical forests was followed by industrialization. Wherever nature's bounty became too hard to secure with ease, Europeans began to steal and domesticate their prizes, and transport them around their empires. The botanical garden at Kew in southwest London became, under Joseph Hooker, an imperial clearing house for exotic species. It established plantations of West African oil palm in India, Jamaica and Australia; sent the macadamia nut from Queensland to the West Indies, South Africa and Singapore; distributed the South American pineapple and Australian eucalyptus tree around the world; and transplanted tea from the forests of the Far East to India and Sri Lanka. Vanilla and cacao beans were snatched from Mexico and grown in British outposts in West Africa.

Britain's India Office recruited several famous Victorian botanists, including liverwort fancier Richard Spruce, to steal cinchona seeds from the forests of the Andes and establish plantations in Sri Lanka and India, where its bark could fight the growing scourge of malaria among its officers and administrators. Most famously, and profitably, in 1876 British explorer Henry Wickham smuggled seeds of Amazon rubber trees from Brazil to Kew, which cultivated seedlings that formed the basis of giant plantations in Malaysia, thus undercutting Brazilian and Belgian suppliers, who were taking latex from the wild.

By the 1920s, British Malaya produced three-quarters of the world's rubber and ran a price cartel. To break it, the French company Michelin established its own massive plantations in Vietnam. Meanwhile, American carmaker Henry Ford set up Fordlândia, a plantation in the Amazon to make his tires.[5] Fordlândia succumbed to leaf blight, but his friend Harvey Firestone was more successful in Liberia, the state established in West Africa to take returning American slaves. In 1926, Firestone secured a ninety-nine-year lease to take over four percent of the country's land to dig up the natural trees and replace them with a rubber plantation. He paid a derisory four cents per acre. The Firestone plantation at one point accounted for forty percent of the country's economy, but was abandoned to squatters, bandits and charcoal burners during the civil war from 1980 to 2003.[6] When I visited in 2010, American managers were back in charge of the plantation. The trees were being renewed and the Liberian government had extended the lease till 2040. The rent had risen to two dollars per acre.

I found the plantation astonishing. As I passed through the gates, I left behind a world of roadside shacks, potholes and bush for what felt briefly like suburban America. There was a golf course, a Mormon church, company schools, a hospital and a radio station. Ancient yellow American school buses drove down roads with signposted junctions and mowed grass verges, transporting workers from their compounds to their trees. The illusion was only broken when I saw the workers' compounds. The housing comprised rudimentary, four-room shacks with no plumbing, just common latrines and two taps for five hundred people.

Production methods were similarly unreconstructed. The seven thousand workers were each allotted 750 trees to tap. They carried the latex to weigh stations in buckets hung over their shoulders. After a centrifuge dried the latex, it was trucked through the gates to a company port, where company ships took it for processing in the US. Not a single rubber band, condom or tire has ever been made in Liberia. Even the old rubber trees that no longer produced latex were being shipped out. As I left the plantation, I drove down the road behind trucks taking them to port for shipping to Europe. They were earmarked to be burned in power stations run by the Swedish state energy company Vattenfall, even though the Liberia outside the plantation remained chronically short of energy. Back at my hotel in the capital, Monrovia, there was another blackout.

HUMANS HAVE CUT TREES for thousands of years. To provide wood for an ever-growing wish list of products that in modern times range from chopsticks to wooden pallets, toilet seats to paper, oboes to flooring, boardwalks to fruit bowls and scaffolding to garden furniture. Increasingly too we have cleared forest ground for growing crops, grazing animals and building settlements. In the past two centuries, our ever-increasing populations and technical prowess have pushed the process into overdrive.

The woodchoppers' heyday in Europe was between 1750 and 1850, when the forested area of the continent west of Russia was reduced by almost fifty million acres, an area the size of England and Scotland. In North America, it was much the same, but nobody was counting. After that, both Europeans and North Americans did much of their logging in other people's countries and there has been some recovery in the forests at home. Meanwhile, the epicenter of deforestation and destruction moved to the tropics.

Huge areas of West African forests were cleared in the late nineteenth and early twentieth centuries. The story of deforestation in the British colony of Sierra Leone, a colony established by Britain to receive freed slaves, was typical. "Most of Sierra Leone was formerly covered by evergreen and semi-deciduous forest," wrote Vivien Gornitz of Columbia University. "In the early nineteenth century, logging for teak and other hardwoods began along the rivers of northern Sierra Leone, and expanded southwards." By the mid-twentieth century, she estimates, only three percent of canopy cover remained, with whole landscapes given over to plantations of coffee and cocoa. "The mountains were bare and the plains largely grassland, with only patches of secondary thickets."[7]

Much of Central America was also cleared of trees a century ago, often by rapacious American corporations such as United Fruit seeking lands for banana and other fruit plantations.[8] Malaysian and Vietnamese rainforests were razed to make room for rubber.

Along the way, we invented the chainsaw, one of mankind's most unsung but ecologically consequential ideas. Originally devised by medics as a small device to cut bones, the chainsaw was reinvented as a machine big enough to fell trees by Samuel Bens of San Francisco in 1905. He wanted to cut down giant redwoods, and axes and saws just took too long. The early chainsaws were too cumbersome for regular use by foresters. It was only in the 1950s that manufacture with aluminum made them light enough for one man to carry. Sales took off immediately. Today, an estimated thirty million chainsaws are bought every year. Their harsh sound has replaced birdsong in most of the world's forests. They cut tens of billions of trees a year. Where once we made forest clearings, now—thanks to chainsaws, together with bulldozers to remove the felled timber—we can clear whole forests.

The chainsaw played an increasing role in the rapid deforestation of successive countries in East and Southeast Asia after the Second World War. The biggest market then was the booming economy of Japan. Its loggers moved from the Philippines to Thailand and Malaysia to meet an insatiable demand back home. Each country in turn lost most of its rainforests, before the loggers moved to the richest pickings of all, in Indonesia.

If producing timber is the only goal, loggers of tropical rainforests usually cut selectively, taking the most valuable species and leaving the rest. Increasingly, however, the aim is to clear the trees permanently, to make way for cattle pastures, peasant smallholdings or, most profitably, plantation crops. In the Brazilian Amazon those crops have primarily been soy, while in Indonesia it is oil palm.

Forests in most developing countries are the final frontier, where the poor and dispossessed, the rich and omnipotent, or the plain criminal can all attempt to fulfil their dreams by felling trees and colonizing the land. They are also traditionally where governments have sent their excess, inconvenient or troublesome citizens. Indonesia's Transmigration Program moved millions of poor people from densely populated Java to outlying, emptier forested islands. Brazil established almost two thousand settlements for migrants across the Amazon after 1970.[9] Typically, the farmers were left to fend for themselves, by clearing forest to grow crops and raise livestock. As a result, they often get blamed for the deforestation. There is a similar game of blame-deflecting in Peru. The government there told a UN climate conference that "ninety percent of logging and burning of Peru's Amazon forests occurs at the hands of peasants who migrate from the highlands," having previously set up policies that actively encouraged poor highland people to migrate to lowland forests.[10]

Colonizing large forests requires central planning and the kind of infrastructure that only governments can provide. For penetrating the rainforests requires access, and that means roads. They allow people in—loggers and hunters, smallholders and gold miners, migrants and refugees, criminal gangs and narco-traders. And they allow the forest's riches out. Along the way, they make the land more valuable for growing traded commodities such as soy and beef. Bill Laurance of James Cook University says ninety-five percent of forest destruction in the Amazon has occurred within three miles of a road.[11] Most governments want roads both for economic "development" and to ensure their writ runs. Few governments these days can rule without roads, whether to bring in their officials' four-by-fours, their police vehicles or, if necessary, their military hardware.

The clearing of the Brazilian Amazon began in earnest in the 1970s, when federal roads allowed the government to truck in landless people from its eastern cities to create new farming colonies.

The BR-163 from Mato Grosso in the south to Santarém in the northeast opened up the "arc of deforestation" along the Amazon's southern and eastern flank. It was completed in 1976, but only fully paved in 2019. In the Brasília bus station that year, I noticed there were new regular bus services all the way to Santarém, a journey of more than two days. Veteran Brazilian environmental scientist Enéas Salati once said that the best way to save the Amazon would be to blow up its roads.[12]

Conservationists say the biggest threat to the Brazilian rainforest today is President Bolsonaro's project, launched in mid-2020, to pave the BR-319 highway.[13] It was built by the military in the 1970s, linking Porto Velho, near the border with Bolivia, to the jungle capital of Manaus. It soon fell into disrepair and is now usable for only half the year. Bolsonaro also talks of reviving an old military project to extend the BR-163 north from Santarém to the Suriname border, and to pave the Trans-Amazonian Highway, which is more pothole than road for most of its fourteen-hundred-mile route east–west. Such projects will "open the floodgates" to loggers, miners and land grabbers, says Philip Fearnside, a veteran American-born forest scientist who has for many years been based at the National Institute of Amazonian Research in Manaus.

Thaís Vilela of the California-based Conservation Strategy Fund agrees. Evaluating the likely impact of seventy-five road projects in the Amazon planned by Bolivia, Brazil, Colombia, Ecuador and Peru, he estimates that they will deforest six million acres by 2040, much of it in protected areas.[14] Paving the Trans-Amazonian Highway alone would doom more than a million acres of forest.

It is abundantly clear that keeping roads away from them is a very good way of preserving forests. There is a single seventy-mile gap in the Pan-American Highway, which otherwise extends for twenty thousand miles from Alaska to Tierra del Fuego. That gap, on the border between Panama and Colombia, contains the Darién swamp forest, one of the last rainforests in Central America. The forest has survived ranchers pushing up to its borders and drug couriers taking boats through the swamp. But WWF lists Darién as one of the world's top ten most threatened forests, and few doubt that the completion of the highway will seal its fate.

Roads imperil forest wildlife, especially the large iconic species that need uninterrupted terrain. In Asia, an analysis of the habitat of the world's last four thousand tigers, much of which is forest, found that eighty thousand miles of roads currently run through recognized "tiger conservation landscapes" in thirteen countries. More than half the animal's habitat is within three miles of a road and there are plans for a further fifteen hundred miles of roads by 2050.[15]

In Indonesia, the proposed Trans-Sumatra Highway would blast through what remains of the island's orangutan forests. The Trans-Papua Highway would pave 2,500 miles of roads across the Indonesian half of the island of New Guinea, "slicing through some of the largest intact tropical rainforests in the world," says Laurance. The three-thousand-mile Pan-Borneo Highway would link the Indonesian, Malaysian and Bruneian parts of the island while being, he says, "a nightmare for endangered species such as the Bornean orangutan, clouded leopard and dwarf elephant."[16]

Africa still has a remarkably low density of roads—just a fifth of the global average—and only a quarter of them are paved. All this could change, however. The African Union has a plan to increase the length of roads on the continent sixfold, creating a network of paved arteries across the continent.[17] The Trans-African Highway project would finally complete the north–south "Cape to Cairo route," once the obsession of British imperialist Cecil Rhodes, as well as linking west to east from Dakar in Senegal to Djibouti. China is helping. All across Africa, I have seen teams of Chinese road builders working at breakneck pace to lay roads, often as part of government-to-government deals that grant China access to African minerals.

Without regular maintenance, many of these roads may not survive the rigors of the rainforest. In the Congo basin, the theoretical length of roads has risen by sixty percent in the past twenty years, doubling in areas where loggers are at work. But, with few other inhabitants, almost half these logging roads soon become overgrown when the loggers move on. This led Laurance and his co-authors to conclude that deliberately decommissioning logging roads after use "could play a crucial role in reducing negative impacts of timber extraction on forest ecosystems."[18]

In the real world, it is not always possible to avoid the presence of roads, or desirable to abandon them back to the forest. The human benefits of being able to get in and out of these landscapes are undeniable. Roads allow farmers to bring their produce to market, and teachers and nurses to work in rural schools and clinics. In surveys, villagers say that roads top their list of desired development projects. So can roads and nature coexist? This is perhaps the biggest challenge for advocates of sustainable development and we will return to the question later in the book.

— 8 —

# LOGGED OUT

## *Well, Almost... Three Decades in Borneo*

BACK IN THE 1990S, I was in central Kalimantan, the most remote province of Borneo, visiting the Sebangau forest. The largest lowland forest on the island grows in waterlogged peat swamp. The Indonesian government had designated it as a protected area, because of its population of orangutans. Two years before my visit, much of it had been set aside as a "living laboratory" to study the natural processes of the swamp.[1] I was the guest of the laboratory's chief researcher, British peatland geographer Jack Rieley of the University of Nottingham.

It didn't take long to realize the limits of the protection offered by the Indonesian government. As we waded almost knee-deep in water, we saw people ahead. A dozen young men were paddling three canoes down a canal they had evidently cut through the swamp. They had their lunch in plastic bags, but in the bottom of one canoe were two chainsaws. Rieley's colleague Nicola Waldes and I watched as they stopped close to a thick red log. It was meranti, an endangered species that made good plywood. There wasn't much doubt what their plans were.

We approached. The gang's leader was brazen. He gave us his name: Udin. His group were Banjar people from the south of the island. They had been brought to the forest three months before to cut trees. They had chopped down the meranti the day before and were now returning to take it to a nearby sawmill. No, he did not know the area was officially protected for science and orangutans. In any case, he shrugged, he was just doing his boss's bidding. We asked who his boss was. He told us that too, naming a Javanese entrepreneur who lived in Kereng Bangkirai, a village just outside the reserve. The canal had been dug on his orders to get the big logs out.

It was all friendly enough, but Waldes asked me to put down my camera. "It is dangerous to confront these people. They have police protection," she said. So we chatted with the rainforest plunderers and ate our respective lunches. They did not push their luck either. The chainsaws stayed stowed and the meranti log was left behind as they paddled away. No doubt they would be back later.

There was worse to come, however. Waldes was taking me to see the most treasured corner of the 125,000-acre laboratory, where she had assembled samples of all the local trees and swamp plants. When we arrived, we found it wrecked, with broken wood and toppled trunks everywhere. Metal tags that once labeled each sapling were strewn around.

Waldes was distraught. Months of work lay wasted. She figured Udin's gang must be to blame, though they were just bit players in a far bigger game. There was, she said, a well-organized illegal operation going on across the lowlands of Indonesian Borneo to extract valuable timber. The crime bosses behind it were bringing in workers from outside the area to do the logging. Corruption was undermining law enforcement. There was enough money sloshing around to butter up the police, pay off inquisitive judges and bribe voters at election time. Even small-time local officials were paid handsomely to turn a blind eye to activities in forests under their control.

Until recently, the Sebangau swamp forest had been left alone by the gangs, because it was hard to penetrate, Waldes said. But with other forests being logged out, its value was rising. So loggers such as Udin were digging canals and laying ramshackle wooden rails over the swamp to get the timber out. Later, just outside the reserve, we saw thousands of logs lashed together and floating on the Sebangau River, waiting to be loaded into the sawmill on the bank. It was all illegal and all in the open. Back in my hotel room later, I noticed that the local phone book listed thirty-one sawmills.

My journey to Sebangau came soon after the fall of Indonesia's longtime dictator Suharto. Many had seen him and his business crony Mohamad "Bob" Hasan as the masterminds behind the growing carnage in Indonesian forests. But it was already becoming clear that Suharto's demise had simply unleashed a new phase of illegal logging in which there were no masterminds, just a great deal of criminality. As political power in Jakarta imploded, local mafias were taking charge of the forests. One Chinese boss of a legal timber concession complained to me that pirate loggers had burned down his local headquarters, because he had asked villagers to report invasions by illegal loggers. "Their organizers are untouchable. Many of them are army people," he said.

Five years later, I was in touch with Rieley again. With support from students at the University of Palangka Raya, his group had managed to stop logging in the reserve. Two decades on, much of the forest remained. It was a triumph of sorts, though the logging syndicates are still in business. They have just moved elsewhere. Like Udin on that morning in the swamp, they are pragmatic, and when the law shows up, they ship out.

⸙

I SAW THE RESULTS of illegal logging in 2007 when I visited Riau, an Indonesian province on the nearby island of Sumatra.[2] I was taking a boat along the Indragiri River, snaking through the rainforest in the company of local environmentalists. We heard the occasional sound of a motorbike in the distance. There was a mobile phone mast behind some trees. Yet all seemed peaceful. The riverbanks were still forested on either side as our boat entered the community of Kuala Cenaku. But as I clambered off the boat and walked through a dilapidated riverside shack, it felt like I was entering a new world. For, behind the shack, the trees were replaced by an empty, mangled and burnt land. I had gone from the old Sumatra, an island until recently almost entirely covered in rainforest, into the new Sumatra, a rapacious world seemingly devoted to devouring that forest.

Our host was Mursyid Muhammad Ali, the head of the seven-thousand-strong community. He was angry at what had happened there just a few months before, when loggers arrived uninvited and unannounced. Till then, he said, the community had managed the forests all around with care. They were communally owned and provided most of their needs. His people harvested rattan creepers to make furniture, took honey from beehives, cut branches to make and repair their homes, and planted a few rubber trees in the clearings to make cash from the latex.

"Then one day, we were just robbed of our land," he told me. Some loggers came up the river, unloaded their chainsaws and told the villagers that they had a permit. The land had been given to them by the government. They began bulldozing the bush. They dug a canal to drain the peaty soil and took their chainsaws to the trees. Later, they set fire to the leftover scrub. When they departed, there were no trees for a distance of three miles from the river.

It was a gross violation of the land rights of the villagers. On the day the loggers had arrived, Mursyid had complained to the district council. "It said it would issue a warrant for the company to stop. The company ignored that and I have had no response since," he told me. In truth, nobody was interested in the fine print of the law, or the rights of the villagers. A year on, it was too late anyhow. The forest was gone. "We have no means of living here now," Mursyid said. "People are leaving the village to find jobs." For them, it was the end of the old Sumatra.

Local investigators, known as the Eyes on the Forest, later told me the loggers were from a subsidiary of the Duta Palma Group, which was largely owned by military men. The company had cleared 25,000 acres of forest in Riau in the previous eighteen months. The most valuable trees, such as ramin, were taken to a sawmill. The rest were chipped and then trucked to a giant pulp mill built some forty miles away in the middle of former dense forest. The pulp mill was one of the world's biggest and was run by one of the world's biggest paper companies, Asia Pulp and Paper (AP&P). The company was gobbling up the Sumatra rainforest as fast as loggers could get the wood to the mill. In many areas, it had cut its own logging roads through the forest. Giant forty-four-wheel "road trains" raced down them, bringing around twenty thousand tons of timber a day to the mill. And as if that wasn't bad enough, there was another similar-sized pulp mill just down the road run by AP&P's big rival, APRIL.

Against that kind of industrial might, the people of Kuala Cenaku stood little chance. Paper from these mills is sold in almost every country around the world. In the early years of the twenty-first century, nowhere in the world was being deforested faster than Indonesia. And nowhere in Indonesia was being cleared of its trees faster than Riau, home of the giant mills turning the forests into shiny white paper to fill desktop printers across the planet. Probably mine included.

All the while, the two pulp behemoths claimed scrupulous legality. Maybe so. The same could not be said for many of their suppliers, however. Generally, they had permits to log, but that was not the same as making the logging legal. Seven years after my visit, the by-then-former governor of Riau was jailed for illegally supplying permits to, among others, companies supplying AP&P's mill.[3] More than twelve million acres of the province's forests had gone.

As the forests have been pulped, the companies have begun planting mile after mile of fast-growing acacia trees to replace them as feedstock for their mills. The new plantations are harvested every five or six years. They called this "sustainable," but to see one of the most complex ecosystems on the planet replaced by endless tree monocultures is to witness an ecological holocaust.

Alongside the new acacia plantations, there are also vast areas of deforested land turned over to the cultivation of oil palm—again to feed voracious global and local markets. That is what happened to most of the land around Kuala Cenaku, which was cultivated by PT Bayas Biofuels, a palm oil company with a mill on the river close by. In late 2019, the company was contracted by the Indonesian government to convert its palm oil into biodiesel as part of its efforts to reduce greenhouse gas emissions. This is a climatic as well as an ecological travesty. The lost forest and drained peat bog on which the oil palm now grows will for many decades have leaked far more greenhouse gases into the air than will be saved by the new "green" fuel.

Nowhere in the world, not even the Brazilian Amazon, has seen deforestation on the scale that has happened in Indonesia in recent decades. Many of the world's most complex ecosystems, evolved over millions of years, have gone. To drive through such landscapes and see mile upon mile of former jungle reduced to bare earth is jaw-dropping. The worst of the deforestation in Indonesia may now be over. Much of the most accessible and valuable forest is gone. The government of Joko Widodo, the reforming president elected in 2014, imposed a moratorium on new licenses for expanding farming into primary forest. Its enforcement was fitful at first. But the average loss of primary forest peaked two years later and has fallen sharply since.

The moratorium was made permanent in 2019. By then, Indonesia was down from first to third place in the league table of annual loss of primary forests, behind Brazil and the Democratic Republic of the Congo.[4] The threats had not disappeared and there was an uptick in rates of forest loss during the coronavirus lockdown in early 2020, when forest patrols were suspended. Even so, with Brazil under President Bolsonaro at least temporarily back among the bad guys, Widodo's turnaround had been a notable achievement.

⁂

WELL, I SAY NOWHERE has suffered worse deforestation than Indonesia. There is one place: the Malaysian state of Sarawak, which occupies a long strip on the north side of Borneo. The carnage here was overseen by one man, the state's all-powerful chief minister Abdul Taib Mahmud. For most of his thirty-four-year rule, which ended in 2014, he ensured that its rainforests were turned into logs by a handful of giant timber companies, many controlled by people with whom he had close ties. Sarawak, a state almost as big as England but with a population then of just 1.5 million, became the world's largest supplier of logs.[5] Its meranti and keruing hardwoods used for furniture, paneling and plywood fetched top dollar. By the time Taib stepped down, just five percent of the state's primary forests remained.[6]

I have visited Sarawak several times over the years. At the height of the destruction in 1994, I was taken on a "facility trip" to show off to journalists the "sustainability" of the logging. The trip was organized by a top London PR company and paid for by the Malaysian Timber Council. The experience was surreal. We were put up at a Hilton Longhouse in Batang Ai, "Borneo's number one nature retreat." The lodge and its grounds were a small enclave surrounded by loggers. It sat beside a scenic lake that turned out to be a hydroelectric reservoir which had flooded the land of the local Iban people. Once known as the wild men of Borneo, the Iban now worked as chambermaids and waiters on the longhouse veranda.[7]

The notorious chief minister joined us. He had just announced plans to commandeer more "idle" Iban land for logging. I asked him about his conservation record. "Of course we protect our forests," he told me. "People here love nature." He was there because local botanists had named two recently discovered shrubs after him and his wife.

Later, our hosts took us to see a plywood factory nearby, run by Samling, one of the largest logging companies in the state. The packs of plywood and veneer stacked up in the warehouse had stickers saying they were manufactured "from sustainable resources." "Oh, yes, we are sure of this," the jovial Japanese manager told me. "All Sarawak's timber is sustainable." It was nonsense. Even the simple rules were not followed. The rules said trees with a diameter of less than two feet should not be cut, but many of the trees in his yard were smaller. The state's own statistics recorded that the timber companies harvested more than twice as much timber as its own advisers said nature would replace.[8]

Flying later along the route of the Baram River, upstream of the Samling factory, I wrote in my notebook: "At almost every bend there seems to be a timber yard. Everywhere the hills are marked by great yellow welts of skid tracks and roads, where logs are dragged from the forest to be driven away." The truth was that nobody seriously believed the hogwash about sustainable forestry. A World Bank report around that time said that logging in Sarawak was a "sunset industry," because the trees were running out. In any event, much of the logging was done in areas being cleared for oil palm. A great deal of that land, it later emerged, was sold on the cheap to companies directly connected to the chief minister himself.[9]

In late 2016, I was back in Sarawak and saw the final phase of this transformation from rainforest state to palm oil plantation. My visit began with a twenty-minute flight from the port town of Miri to the small inland town of Marudi. It was a flight I had taken twenty-five years before. The change in the landscape was staggering. The first time, it was rainforest the whole way. This time the trees were all gone. Almost the entire route was laid out in rectangles filled with oil palm, bordered occasionally by smashed-up forest fragments scattered with abandoned logging equipment.

I had returned to Sarawak to discover who had killed a political activist attempting to halt the tide of palm oil. One June morning earlier that year, Bill Kayong was fifteen minutes into his commute to Miri. He was waiting in his pickup truck at traffic lights across from a shopping mall when two bullets shattered the side window and struck him in the head. Five months later, I made my own journey into town to watch as three Miri men stood in the magistrates' court charged with his murder. They were a nightclub bouncer, a karaoke bar operator and a man described as the personal assistant of Stephen Lee, the head of a Malaysian palm oil company called Tung Huat. Kayong had been in conflict with this company, owned by Lee and his father, both members of the ethnically Chinese Miri business community.

Later, I met Kayong's boss, Michael Teo, an energetic local doctor and a leader of the main Sarawak opposition party, the People's Justice Party. Teo had himself been beaten with a baseball bat near his clinic and left with a collarbone broken in three places. "I was told I would be killed because I was involved with Bill," he said as we spoke at the café where it happened. "Then one morning someone phoned me and said Bill had been killed."

After his death, some in the media described Kayong as a "dedicated environmentalist." That's not quite right. Unlike Bruno Manser, a Swiss environmentalist who disappeared without trace in the Sarawak forests in 2000, Kayong was primarily a political activist. He was dedicated to protecting the rights of native communities in Sarawak, known as Dayak, from the increasingly strong-arm invasions of their remaining traditional forest lands by logging and palm oil companies. Lately, Kayong's work had been concentrated on helping one community south of Miri in their battle with the Lees.

The people of Sungai Bekelit lived in a traditional longhouse, a large wooden building some three hundred feet in length, raised on stilts and with lines of apartments off a wide, covered communal area. Like others, this longhouse was a legally established social unit, with communal lands held under customary laws that dated back many centuries. Many environmentalists see longhouse communities as the last hope for the state's dwindling forests.

They are certainly tenacious. For eight years, the Sungai Bekelit community had been fighting the Lees' attempted takeover of their partially forested land. The community said their customary rights took precedence. That should have been true. International norms say Indigenous communities have the right to give or withhold their "free, prior and informed consent" to economic activities on their traditional lands. But officials in the Sarawak government, who granted the Lees a license to the land, claim never to have heard of such norms.

Longhouse living is hardly idyllic. In remote upcountry regions reachable only by boat, conditions are hard. But nearer to roads, the residents often have pickup trucks, satellite TV dishes and mobile phones. Some have jobs in the cities too and come home only at weekends. Even so, they retain fierce attachment to their land, especially when the government tries to sell it off. "Before the logging company came, life was easy because we had the forest," said one headman at a longhouse that had tried to reforest part of its land. "We used to hunt wild animals. The water was completely clean. But it's more difficult to feed our families now."

The Sungai Bekelit longhouse held three hundred people in sixty-three families. When I visited, one of the women took me to see a blockade that the longhouse residents manned twenty-four hours a day to prevent the Lees' henchmen from taking their land. "The Lees were very scared of Bill," she said. "He was becoming powerful and they wanted to silence him." A man from a neighboring longhouse told me, "They offered to pay him to control us. When he recorded that conversation and played it to us, they accused him of treachery. Then it got ugly."

In the big scheme of things in Sarawak, what was happening at Sungai Bekelit was a small dispute over not much land. Most of the rich pickings for the unscrupulous had long since been taken. What forest remained was being scrapped over by small-time racketeers with more guns than sense. There were no international headlines when Kayong was shot, or public shrines to his life. He was buried in a modest cemetery outside Miri. By the time of my visit, the tiny commemorative photograph of him pinned to a pole on the roadside where he died was frayed and overgrown by grass.

At the court hearing I attended, a new Dayak grassroots membership organization that Kayong helped found had turned out in force. A dozen of its young members sat on benches wearing black T-shirts in his memory. They included his three brothers. Outside in the corridor, his wife was comforted by their two teenage children. "I didn't know about any threat to him," she told me. "He kept it to himself. He didn't want us to be worried."

In court, it was clear the police believed Stephen Lee was the "mastermind" and they thought that he and his assistant had hired the other two to carry out the killing.[10] Lee was on the run, but the other three were remanded for a later hearing. That afternoon one of them, the nightclub bouncer, Mohamad Fitri Pauzi, was back in the dock, accused with two other accomplices of another attack. This time, it was on the Sungai Bekelit longhouse's headman, Jambai Anak Jali, who had been working with Kayong.

Jambai, a mild-mannered, neatly dressed man in late middle age, described to the court how his car was followed, rammed, forced off the road and turned over. As he, his wife and their niece lay inside, the men beat the windscreen with baseball bats and slashed him with a samurai sword. It was, he told the court, the latest in a string of assaults on him, his house and his car since 2008, when he had first gone to court to contest Tung Huat's license to cultivate forest land around his and four other local longhouses. "We can prove it is our land. We have been there since 1934," he told the court. The shaven-headed Fitri stood impassively in the dock and again pleaded not guilty.

Back home, I continued to follow the Kayong case. The police eventually caught up with Lee after a manhunt that began in Singapore, moved to Melbourne and ended in the Chinese province of Fujian. At the full trial the following March, Fitri was convicted of the shooting and in August 2018 was sentenced to death, though at the time of writing that execution had not been carried out.[11] Lee and the others were acquitted. "The mastermind managed to get away with it," said Kayong's brother Davey. There was worse. In October 2019, Jambai and the longhouses lost their claim to their land. A federal court in the state capital, Kuching, ruled that customary law did not take precedence over the government lease to the Lees. This was widely regarded as a landmark ruling. It was a very depressing one for both the Dayak communities and the remaining forest.

— 9 —

# CONSUMING THE FORESTS

## *Logs of War and a New "Green" Plunder*

THROUGH THE 1990S, the dirt-poor, socially divided African state of Liberia suffered a long civil war, during which a quarter of a million people died and half the population fled. The war had many origins. They included a botched coup and constant tension between native locals and the descendants of former African slaves, for whom the state was created in 1847 and who tended to rule things. What sustained the conflict, however, was the flood of arms that kept the warring factions in business. Most of the money for these arms came from selling rights to log the country's rainforests, the most extensive in West Africa. For years, as battles raged, ships regularly left the ports of Monrovia and Buchanan loaded with what came to be known as the logs of war, returning later with their holds full of arms. A UN study later found that eighty-six percent of the country's timber production was at the time controlled by arms traders.

The kingpin of this carnage was Charles Taylor, who began the war when he mounted an armed invasion. Even before he seized the capital, Monrovia, he was selling logging rights to land he controlled. Once installed as president, he turned his criminality into a state enterprise, though most of the checks paid for logging rights were made out to his personal account. He chose as his partners some unlikely foresters. One was Leonid Minin, a reputed Ukrainian mafia boss. His Exotic Tropic Timber Enterprises supplied Taylor with arms, in return for a license to log the Cavalla Forest in the northeast of the country.

Another was Guus Kouwenhoven, a Dutch adventurer once famous for trading in stolen Rembrandts, and who ran Monrovia's wartime casino before Taylor gave him a government job issuing logging licenses. "Mr. Guus," as he was known, awarded his own company, Oriental Timber, the logging rights to over a quarter of the country's forests. He shipped in Asian workers to cut the trees and Asian prostitutes to service the workers. Part of the proceeds from the sales of the timber to customers in France and China bought guns for Taylor.

The tragedy was that the world knew all along what was going on. A British diplomat told journalists in 2001 that "it is the timber trade that is keeping Taylor in power."[1] But the international community failed to act until May 2003, when the UN Security Council finally imposed a trade embargo. The effect was instant. By August, the regime had collapsed. After sixteen years, the war was over.

Taylor was eventually convicted of war crimes by the International Criminal Court in The Hague and sentenced to fifty years' imprisonment. After over a decade of tortuous legal process, Kouwenhoven was sentenced by a Dutch court to seventeen years for arms trafficking.[2] He refused to attend the hearings, however, and stayed holed up in South Africa. There, in early 2020, a magistrate ruled that "with great regret," he could not under South African law extradite him to undertake his sentence.[3] Minin also remained at large and reputedly still trading arms.

Liberia's forests, though depleted, have shown a great ability to recover. Even so, what remains staggering is how long the logs of war were allowed to fund a vicious conflict that an ounce of probity in the international timber trade could have halted. Illegality, it seems, has been second nature to the industry.

According to deforestation analyst Sam Lawson, founder of the Earthsight investigative NGO, more than 3.5 billion cubic feet of timber was logged illegally from the world's forests in the first decade of this century.[4] That was a tenth of the total harvest and enough logs to stretch ten times around the Earth. Other international bodies, such as the European Union, put the figure as high as forty percent.[5] It probably depended on how you defined illegal. In some countries there has been more illegal logging than legal logging, with criminals damaging or destroying around ten million acres of forests each year.

When not funding wars, the international trade in illegal logs has often been propping up tyrants. Cambodian tycoon Oknha Try Pheap, often known as "the king of rosewood," has in recent years been widely accused of masterminding illegal logging with the connivance of the country's longtime prime minister, Hun Sen. In 2016, I interviewed the man who exposed the scandal, Cambodian lawyer Leng Ouch, after he won a Goldman Prize, awarded annually to grassroots environmental activists.[6] The son of poor farmers, Ouch had infiltrated logging syndicates, working undercover as a laborer, timber trader and even a cook to document how Pheap plundered trees from the country's national parks.

Most of the Cambodian timber went to Vietnam for shipping to China, where rosewood is made into antique-style furniture, known as *hongmu*, which is prized by the Chinese nouveau riche. Studies show that Cambodia has had the fifth-fastest rate of forest loss in the world. Ouch said the government in effect works for Pheap: "He can get rid of provincial governors. The military are deep in the logging business." At the end of 2019, the US Treasury agreed.

Imposing sanctions on Pheap, it said that he "used his vast network inside Cambodia to build a large illegal logging consortium that relies on the collusion of Cambodian officials... including military protection."[7] Hun Sen's government dismissed the allegations as "groundless."

⁂

LIKE MUCH THAT CHINA has done in the world, its dominance of the global timber market is recent and huge. It began in 1998, when Beijing banned logging in its own natural forests after widespread floods on the Yangtze River that year were blamed by its scientists on deforestation in the river's headwaters. A couple of years later, I saw myself how effective the ban was when I drove through the mountains of northern Sichuan. All along the road, timber-processing factories stood empty. Roads were deserted. "Until two years ago, outsiders were banned from many roads in northern Sichuan because they were used exclusively for logging vehicles," said Chen Youping, a local forestry official.

To sustain its fast-growing timber-processing industry, China erupted onto the international timber market. Since the turn of the millennium, Chinese timber companies have often taken over from European colonial companies in logging the forests of West and Central Africa. Overnight, its demands unleashed a major black market in illegally harvested timber from the Siberian forests around remote cities like Irkutsk and Tomsk. In the early days, supply was organized by mafias in "bandit villages," often using former Soviet military vehicles. It was dangerous to object. Wood processor Vladimir Baranov, who refused to supply Chinese buyers, was shot dead on his doorstep in 2005. The illegality has continued. Some estimates say the black market controls a quarter of Russia's timber exports to China.[8] Legal or not, hundreds of thousands of railway carriages, carrying some 350 million cubic feet of timber, annually cross the border from Zabaikalsk, a giant Russian timber yard east of Lake Baikal, to Manzhouli, a once-desolate border town that is now a gleaming Chinese metropolis of a quarter of a million people.

A similar sense of illegality has hung over hardwood imports to China from Southeast Asia and West Africa. At one point, half of all internationally traded tropical hardwood went to China, and two-thirds of that was unloaded in Zhangjiagang, a timber port on the Yangtze estuary. When I visited, even big traders didn't bother with calling cards or offices. They were holed up in anonymous hotel rooms and you reached them by calling the mobile phone numbers scrawled on logs with a marker pen. That didn't suggest a high level of probity.

According to Sun Xiufang, a timber trade analyst now at Virginia Polytechnic Institute and State University, most traders there neither knew nor cared whether their timber was cut legally.[9] They knew what they liked. They were buying okoumé from West Africa, meranti from Borneo, teak from Myanmar, greenheart from Guyana, rosewood from Ghana and merbau from Indonesian New Guinea, almost the last stronghold of a tree highly prized for its density, strength and durability. According to the Environmental Investigation Agency (EIA), one log of merbau was imported to China from New Guinea every minute of every working day, in what it called "the world's biggest timber smuggling racket."[10] Much of the merbau went to Nanxun, a sleepy riverside district near Zhangjiagang. Once known for its canals and old buildings, it is now the hardwood flooring capital of the world.

Europe remains a big market for both Chinese timber products and the raw timber culled from the forests by Chinese logging companies. In late 2019, Indra Van Gisbergen, a campaigner I know from the European forest campaign group Fern, was given a tour of the timber docks at Antwerp. The aim was to show her how well the port policed timber imports to ensure compliance with an EU ban on illegally logged timber. But, she wrote later, "It didn't take long before we saw a consignment of timber whose markings stopped us in our tracks."[11] Logs of padauk, a hardwood from Gabon used in furniture, musical instruments and flooring, were marked as shipped by the Chinese logging company Wan Chuan Timber. The company had recently been named by the EIA as "perhaps the worst offender of crimes committed against the forests and the people of Gabon."[12] Fern had released video of a senior company official boasting about how it laundered illegal logs and illegal profits. The company sent timber to Antwerp every two months, yet port officials apparently raised no questions about the consignment's legality.

Illegal logs, and products made from illegal logs, appear still to be widely available in Europe. The EIA has tracked contraband Burmese teak, most of it cut by Chinese operators, into the EU via the Croatian port of Rijeka and a small Croatian trading company called Viator Pula.[13] From there, it was made into decking for many of the luxury yachts sold in the UK. Sam Lawson says the EU has long been "a global leader in the import of commodities stemming from illegally cleared forests."[14] The taint of dodgy dealing persists, even though the EU has for years had laws requiring timber importers to ensure that their supply chains are clean of illegality.

Things may be changing, both in China and among its suppliers. At the start of 2020, something dramatic happened in the global timber trade—something that could prove transformative. China announced plans to alter an old forest law. The new wording would specifically ban the buying, processing or importing of illegally sourced timber. With so much of the world's timber cut from forests in defiance of national laws, it was a big deal. The EIA, a normally skeptical NGO that has probably done more than any other to expose China's reliance on illegal timber, called the announcement "an extraordinary move." It could force China's suppliers in Russia, Africa and Southeast Asia to clean up their industries.[15]

There were early signs that this was already happening. Just as China's crackdown on illegal timber got underway, the boss of a big

Indonesian timber group, Ming Ho of Alco Timber Irian, began a five-year prison sentence after the authorities intercepted almost four hundred containers of illegal merbau on the dockside in Sorong, a fast-growing West Papuan port.[16] It was a rare case of the top man getting convicted for such a crime. Perhaps, just perhaps, it was a sign of things to come.

Of course, a crackdown on illegal logging is not necessarily the same as a crackdown on all logging. Reining in the rampant logging industry will only happen in earnest once demand for its products is also reined in. The danger is that, rather than reducing demand for trees, the world could be about to open up an entirely new market. Perversely, it is being done in the name of generating "green" energy to fight climate change. I was surprised to see the consequences of this new industry first in central Europe—in Slovakia.

HE SAID HE WAS a forest ranger as he leaped out of an unmarked car, shouting and waving a heavy stick in our faces. But he showed no ID. The three of us enjoying the afternoon sun on a short stroll through the beech forests of Poloniny National Park were trespassing, he said. This was a criminal activity. He harangued us for fifteen minutes as we reluctantly retraced our steps to the main road. All the while we could hear the sound of heavy machinery operated by logging companies inside the park. In Slovakia, it seems, loggers are allowed in the protected forests, but walkers who might watch their activities are not.

Slovakia is a small, sparsely populated country, one half of what was once Czechoslovakia. The logging was happening because its government had adopted a plan to generate much of its electricity by burning biomass—that is, trees—in new power stations, whose construction it is heavily subsidizing. The aim is partly to meet EU targets for expanding renewable energy. If anywhere in Europe could cope with more woodcutting to generate energy, I thought, then it would be Slovakia. Almost half the country's land is forested and wood is central to its culture. Earlier that day, I had admired some of its famous UNESCO-listed wooden churches. But judging by my experience in Poloniny National Park, the evidence of stress on Slovakia's forests was everywhere around us—and not just in the demeanor of an overwrought park ranger.

In the Carpathian Mountains of eastern Slovakia, clear-cutting of beech forests was increasing fast. The area around the park, on the border with Ukraine and Poland, had the highest concentration of old forests in the country and was a prime target. Logging was legal, provided that the loggers had a license. That seemed to be no problem. In fact, logging was encouraged, not just by Slovak politicians, but in Brussels too. All along the route into the park, EU infrastructure funds were being spent to widen the roads and improve access to the forests for heavy vehicles.

I traveled with Peter Sabo from a local NGO called WOLF, which campaigns to protect Slovakia's forests. He said the boom in logging to burn in biomass boilers was wrecking the country's forests, especially in the Carpathians. Government statistics were stark. They showed that 350 million cubic feet of wood were being logged each year, while the annual regrowth was only around 210 million cubic feet. With around 120 million cubic feet being burned for energy and heating, this new use represented almost the entire over-harvest. "Behind these numbers lie destroyed natural ecosystems, places where wolves, bears, deer and others had their quiet places," Sabo said. "In Slovakia, wood burning for energy is not a renewable resource."

There are around a dozen biomass-burning power plants scattered across Slovakia. According to the government, their operators should burn only low-grade timber that would not be suitable for other industrial purposes. That means burning the branches and twigs that foresters would otherwise leave on the forest floor to rot, releasing their carbon into the air. If the industry followed the rules, then Bratislava and Brussels would be right to claim that the burning did not add to carbon emissions. The reality was different, said Sabo. This low-grade wood was already mostly used for local heating. So, whatever the rules said, the new power stations were burning higher-grade timber which had been cut specially for them. That was not carbon-neutral.

To prove the point, he took me an hour's drive from Poloniny to the beautiful medieval town of Bardejov in the Beskydy Mountains. Just outside the town was a power station. It burned around 3.5 million cubic feet of wood each year to generate electricity and supply heat to the town. We peered through the fence at a timber yard full of large logs with diameters of up to three feet. These were not logs with "no other potential industrial uses." As we watched, machinery chipped some of the logs and placed them in a pile within feet of the plant's furnace. "The company has been convicted and fined for this," said Sabo. "But the fines are low and they just carry on."

There was no official at the plant to explain. When I later contacted the company, Bioenergy Bardejov, it denied owning the logs in the timber yard. Stanislav Legat, a manager, emailed: "That is not our yard and not our technology. Owner is another company. We only use chips. Logs whom you see on the courtyard is not ours."

Of course, people have always cut wood to heat their homes. The activity I had witnessed in Slovakia, however, was the start of what could become a vast new business based on plundering the world's forests. Timber has become the main focus of the EU's efforts to switch to renewable energy. Biomass burning makes up around two-thirds of what the EU claims is zero-carbon, renewable energy. A startling forty-two percent of EU cut timber is now burned as fuel. Soon it will no longer be small local power stations such as Bardejov burning local forests. It will be giant power plants originally built to burn coal and now switching to burning logs culled from almost anywhere in the world.

ANYONE TRAVELING UP the east side of England, through the flatlands of eastern Yorkshire, will have seen it. With its twelve giant cooling towers, Britain's biggest power station is unmistakable. Drax at times supplies a tenth of the nation's electricity. Once, it burned coal mined from the nearby Selby coalfield. By 2020, it was mostly burning wood pellets imported from the American South. To keep Drax going, more than six million tons of pellets crossed the Atlantic from North Carolina, Louisiana and Mississippi each year. The power station's last remaining coal-burning boiler was due to be converted to burning wood in 2021. Much of the cost of the conversion had been met by generous subsidies for renewable energy from the British government.

The Drax power station is five hundred times larger than Bardejov and sends into the air each year more carbon dioxide than the entire emissions of 124 nations. So you may be surprised to hear that the plant's operators say it is "the largest carbon-saving project in Europe." The British government declares the pellet burning to be carbon-neutral, so none of its emissions turn up in national declarations. They say this is because the trees being burned by Drax will be replaced with new ones on the same land that will soak up the same amount of gas as is being emitted.

The station's operators took me to see their new American forestry business. Near the small "wood town" of Gloster in Amite County, Mississippi, I watched as truckloads of pine cut from local forests were poured into a giant chipper that in turn sent some 250 tons of timber shards every hour into a drier that halved their volume by removing moisture. The processed pellets from Gloster—plus those from Drax's two other pellet mills at Morehouse and LaSalle in Louisiana, and seven more run by Enviva, the world's largest producer of wood pellets—were then trucked south to Baton Rouge. There I later watched a hopper as high as an English church tower pouring pellets into 55,000-ton ships that set out several times a week on a nineteen-day Atlantic crossing to British docksides at Immingham or Liverpool. All just to supply one power station.

Drax claims that its voracious demand is not driving any increase in deforestation in the Deep South. Far from it. "The effect of Drax's arrival has been to help keep the South forested," Dale Greene, dean of forestry at the University of Georgia, told me at Drax's office in Monroe, Louisiana. "If we harvest more, we plant more and there is more carbon in the forest." Mississippi had in recent years added a million acres of forests, many on former cotton fields, a five-percent increase on forest cover in the early 1990s. Drax says its only carbon footprint is the energy used to harvest, process and transport the wood.

American ecologists don't buy this. They say it is a fantasy perpetrated at the expense of their forests and ultimately of the atmosphere itself. In 2017, some two hundred scientists wrote to the EU insisting that "bioenergy from forest biomass is not carbon-neutral." They called for tighter rules to protect forests and their carbon from Drax and dozens of potential imitators.[17]

How can there be such disagreement? It is true that if the trees are replaced like for like, then the carbon that goes up the stack will be reabsorbed by the new trees. However, that is just the start of the carbon accounting problem. In theory, under the terms of the Paris climate agreement, any overall carbon losses from forests supplying power stations should be declared. But how? Drax does not own any of the forests it is burning. It cannot control what happens on the deforested land afterwards. Nor can regulators and carbon counters. They all hope that market forces will do the replanting, but it may not happen. A British government study in 2014 concluded that a worst-case scenario, in which the logged forest land is turned over to agriculture, could see Drax's real carbon dioxide emissions double those caused by burning coal.[18] We may know the answer only decades after the power station is shut. By then it will be too late to call anyone to account.

A further time-lag issue relates to the replanted forests. Replacement forests will take decades to grow big enough to take up all the carbon dioxide emitted by burning the old trees. "During this time, the additional warming will cause changes such as melting glaciers and thawing permafrost," says William Moomaw of Tufts University.[19] The trees that regrow may also absorb much less carbon than those they replace if they are cut down again too soon. This premature harvesting is already happening. As Greene told me, "They plant quick and cut quick." He suggested turnover time was now only twenty-five years.

After touring the Drax mill and watching pellets leave Baton Rouge, I journeyed to another of Drax's main supply areas, where Plum Creek, one of the US's biggest lumber producers, has been buying up local privately owned forests amid the Baptist churches, beat-up family farms and trailer parks of Noxubee County, Mississippi. The trees that loggers once cut here to make pulp for paper-makers now get hauled to Drax to make pellets, said Greg Knight, the local manager. Business seemed to be brisk, and harvesting was intensifying. We were speaking in the shade of an eighteen-year-old yellow pine plantation covering 300,000 acres that he said had already been thinned twice. Clearly Greene's twenty-five-year turnover time had been reduced.

To be fair, despite the intense cultivation, the plantation was not devoid of wildlife. There were songbirds around, as well as squirrels, gopher tortoises and some white-tailed deer, turkey and quail. Knight said the thinning encouraged animals, because it provided sunlight for foraging. "If you do good thinning, you get wildlife and then you sell the hunting licenses," one local smallholder told me.

Well, yes, but then we walked down to a creek. There, away from the bulldozers, the forest was much less intensively managed. The monoculture of pines gave way to a mix of native hardwoods, including cherrybark oak. Some were eighty years old. It was great to see them. But it showed the difference between native and production forests here, and what was being lost. Forest cover may be increasing, but forest biodiversity is in sharp decline.

Scot Quaranda of the Dogwood Alliance, a forest conservation NGO based in North Carolina, has been tracking how ecologically important hardwoods such as oak and sweetgum have been clear-felled from swamplands across that state and loaded onto Drax's boats.[20] Enviva, which supplied the hardwoods, did not deny the charge. But it claimed that it was making productive use of swamp forests that would otherwise be drained for agriculture.

In any event, says Quaranda, as the hardwoods disappear, the more intensive cultivation of pines to meet Drax's demands meant that naturally regenerating genetically diverse pine would soon all be gone. Replaced by new industrial forests full of uniform ranks of fast-growing pine rushed into service to supply a power station on the other side of the ocean. Greene agreed that there was indeed such a switch going on. It may be true that Drax is helping maintain the economics of forestry in the Deep South, and with it perhaps the forest cover. But it is changing the nature of those forests profoundly.

This is just the start. Demand for wood pellets to burn in "carbon-neutral" power stations is taking off. In Europe, which so far has most invested in biomass burning, proposals on the drawing board in early 2020 would convert ten of the continent's largest coal-burning power stations to burning biomass, mostly wood pellets. That would create an annual demand for forty million tons of pellets, cut from forests equivalent to half the size of Germany's Black Forest, according to one NGO study.[21] Meanwhile, South Korea and Japan are also in the market for timber to make wood pellets. The Sumitomo Corporation has signed contracts to buy American pine to supply power plants around Fukushima, site of the crippled nuclear reactor hit by the 2011 tsunami. Russia is also setting up to become a booming supplier.

For centuries, forests have been cut for firewood and charcoal for local needs. Pelletization makes it for the first time economic to traffic wood halfway around the world to burn in power stations. Supplying Drax is still by some way the largest such trade in the world. But it could soon become the model for a new global industry potentially supplying hundreds of such pellet-burning power stations.

Some advocates see such wood-burning power stations becoming not carbon-neutral but carbon-negative. The technological Holy Grail is to add equipment to strip carbon dioxide from stack emissions. The gas would then be pressurized or liquefied and buried out of harm's way—a technology called carbon capture and storage. So, by putting wood burning together with carbon capture and storage, the world could grow forests that suck carbon dioxide from the air, then burn them and bury the emissions underground. They call it BECCS, for bioenergy with carbon capture and storage. In that scenario, the more energy you generate the more carbon you will take out of the air.

The technology for developing carbon capture and storage on the scale required is at an early stage—and funding for the necessary R&D has been fitful at best. But this is no idle pipe dream. In 2020, Drax launched a pilot project for capturing the gas.[22] Its CEO, Will Gardiner, talked of sending the gas down pipes into empty gas fields beneath the North Sea before 2030.

Climate scientists say ways to remove carbon dioxide from the atmosphere are urgently needed to stem global warming before too much damage is done. Reforesting the planet is one—probably the cheapest and biggest—means of doing that. The question is: What kinds of trees? Ecologists are excited by the dream of restoring the planet's great ancient forests in the cause of ecological healing and climate protection. But BECCS could turn that dream into a nightmare. Instead of great natural forests, blooming with biodiversity, we could end up covering the Earth in vast monocultures of fast-growing trees dedicated to saving the climate and generating electricity.

# —10—

# NO-MAN'S-LAND

## *Cattle Kingdoms and the Tyranny of Global Commodities*

LANDLOCKED PARAGUAY IS a bizarre place. Within minutes of taking off from Asunción airport, our six-seater Cessna was flying over dense thorn forests of a kind found nowhere else on Earth, because nowhere else on Earth has a climate so extreme, with burning summer heat, bitterly cold winters and droughts punctuated by floods. The forests occupy the Chaco region of Paraguay, an empty quarter stretching for hundreds of miles on a plain as flat as a tabletop, becoming ever drier as it goes northwest towards Bolivia. Paraguay and Bolivia fought an improbable war here in the 1930s, in which nearly one in thirty Paraguayans died, mostly from thirst. Otherwise much of it has only ever been trodden by members of the local Ayoreo tribes.

The wild Chaco once covered more than 300 million acres—five times the area of the UK. It extended into Bolivia, Brazil and Argentina. But it has been eaten away by the expansion of agriculture. Most of what survives is its hottest, most distinctive and most forbidding section in Paraguay, the last great wilderness in South America. This heartland is a world of odd creatures—giant anteaters, tapirs, maned wolves, the llama-like guanaco, flightless rheas as tall as a man, and ten species of armadillo. The trees are squat and bush-like, covered in vicious thorns, or with bottle-shaped trunks that hold moisture like a camel's hump. A lot are endemic. Nowhere else would have them.

This unique ecosystem is more ancient, more distinctive and far less studied than the Amazon. The strangeness is partly due to a climate that switches between fifty-degree-Celsius summers (about 120°F) and below-freezing winters, and between searing droughts and extensive floods. Such wild fluctuations, along with the thorns, had kept the modern world out until the last two decades, during which the region has seen an invasion of Brazilian ranchers. While their government at home cracked down on clearing the Amazon for ranches, and the Cerrado grasslands where their herds once roamed free were bought up by soybean farmers, the ranchers moved across the border into the Paraguayan Chaco.

Deforestation is now happening at breakneck speed, to the horror of ecologists who have barely scratched the surface of the Chaco's weirdness. Toby Pennington of the Royal Botanic Garden in Edinburgh calls the Chaco a "museum of diversity." At a time when we fear climate change, "it seems especially crazy to be losing species that are obviously incredibly well adapted to extreme climate," he told me.

Half an hour out of Asunción, the thorns cleared and we were flying over ranches carved out of the forest. Most were laid out in 250-acre rectangles, separated by thin strips of trees. Cattle were clearly visible below us. So too were bulldozers clearing more forest. As we flew on, the ranches became bigger. One covered 125,000 acres, containing five hundred rectangles. It had all been cleared in the previous year, said my flight companion, Oscar Rodas of the NGO Guyra Paraguay, who regularly flew over to chart the destruction. Colleagues said he was far better informed about what was going on in the Chaco than the government.

The rectangles were very unlike the seemingly random pattern of deforestation I saw in the Amazon. Clearing here was systematic, carried out by ranching conglomerates rather than speculators. Once cleared, the ground was seeded with African grasses and cattle were soon installed. "The African grasses are much greener. They grow quicker," said Rodas, as we flew on.

The Paraguayan Chaco had, by the end of the 2010s, one of the highest rates of deforestation in the world. Around seventeen hundred acres were being lost every day, according to Guyra Paraguay. Three large landowning companies were said to be taking the lion's share: Yaguareté Porá from Brazil, Itapoti from Paraguay and the Spanish-owned San José Group. Most of the clearance is legal. The government wants to follow neighboring Brazil and supply global markets for beef and other agricultural products such as soy. It approves almost all proposals to clear the forest. As a result, the cattle population in the Chaco has risen to six million and the country has come from nowhere to being one of the world's top ten beef exporters.[1] It reportedly aims to have a cattle herd of twenty million by 2028.[2] So, in a matter of years, the emptiest quarter of one of the most remote and little-noticed countries in the world has become a linchpin for the onward expansion of a global market to feed the world.

Logging is only part of the story of rich nations' complicity in tropical forest destruction. Much of the clearing is primarily to provide land for plantations growing commodity crops such as soy, palm oil, rubber and cocoa, or to raise beef cattle. These crops are usually grown to supply fast-expanding global markets. In other words: us. Most of us eat rainforests in our burgers, wear rainforests in our shoes, feed rainforests into our printers, wash ourselves in rainforest soap, spread rainforests on our bread and even drive our cars on rubber from former rainforests. The land demands of large-scale commercial agriculture account for at least two-thirds of deforestation in the tropics. Despite Europe's claimed high environmental standards, the European Commission itself estimated that consumption in the EU—at least while Britain was still a member—has been responsible for thirty-six percent of all the forest destruction from growing internationally traded commodities.[3] An increasing amount of that is happening in Paraguay.

The Paraguayan government used to pay lip service to protecting the Chaco, said José Luis Casaccia, a former environment minister and chief environmental prosecutor, when we met at Guyra Paraguay's office in the suburbs of Asunción. It required ranchers clearing forest to leave a quarter of the trees. Casaccia said he had lost his ministerial job because he had proposed upping the proportion protected to fifty percent. Instead, under president Horácio Cartes—who is a rancher himself in the Chaco—the government later dropped the requirement altogether.[4]

"The Chaco is a no-man's-land," Casaccia told me. "The ranchers on their huge estates make up their own laws." I asked for his prognosis for the Chaco. "Apocalyptic," he replied. "On current trends, everything that is not protected will be gone. The Chaco will be reduced to desert, with all the species in it lost." A few areas of the Chaco are protected, but state parks and reserves cover not much more than one percent. Conservation, where it exists, is mostly in private hands.

As we flew north, the ranches beneath us were suddenly replaced by forest stretching towards the horizon. "That's the Moonies' land," said Rodas. During the 1990s, Sun Myung Moon's Korea-based Unification Church bought forests and wetlands in Paraguay and across the border in Brazil. It amassed some two million acres to create what Moon called a "kingdom of heaven on Earth."[5] Part of that kingdom had previously been owned by descendants of a swashbuckling Spaniard called Carlos Casado. For a century, they had harvested the quebracho tree, a name derived from the Spanish for "ax-breaker," because of its extremely hard wood. The tree also contains high concentrations of tannin, used for tanning leather. We spotted one of Casado's abandoned logging railways, now being reclaimed by the forest. The quebracho trees on the estate are long gone, but the Moonies' purchase has protected the forest from clearance by ranchers.

We flew on to the ten-thousand-acre Cardozo estate, west of Bahía Negra, a port on the banks of the Paraguay River, which winds through the eastern Chaco. The estate, though much smaller than the Moonies' kingdom, also had intact forest. The landowner had "gone green," said Rodas. Soon after my trip, he sold to conservationists at the UK-based World Land Trust. The trust then reached an agreement with the fifteen hundred Indigenous Ishir fishers who live along the river. They would jointly manage the Cardozo ranch for twenty years, then the Ishir would assume full title, on condition that the forest is maintained.

Relations were not so harmonious between the Ishir and the Moonies, however. The Moonies' territory includes Ishir sacred burial grounds. "The cemeteries are our most precious land," Cándido Martínez, an Ishir leader and councillor in Bahía Negra, told me. "But we are not even allowed to visit them."

Away from the river, most of the Paraguayan Chaco has been occupied only by some three thousand Indigenous Ayoreo people. Those who have not been recruited as ranch hands live in the bush, aloof from outsiders and occasionally entirely uncontacted. The San José Group has been accused of illegally bulldozing forest where uncontacted Ayoreo live.[6] Environmentalists see both the Ishir and the Ayoreo as allies against the ranchers. But scientists have hit problems. In 2010, sixty British botanists and zoologists planned a big expedition to discover new species in the Chovoreca Natural Monument in the remote northwest of the Chaco. Then, weeks before they were due to set off, Ayoreo representatives objected that the scientists might stumble on one of the ten or so families of isolated Ayoreo who live in the park.[7] The trip was canceled, never to be revived. It was a shame, said Alberto Yanosky, the head of Guyra Paraguay. The Ayoreo needed the forests to be protected, and greater outside knowledge of the Chaco's unique ecology might be the best chance of achieving that.

I never got to meet any Ayoreo people. But arguably I did meet the oddest tribe of all in the Chaco. After a bone-shaking seven-hour bus ride up the Trans-Chaco Highway from Asunción, a neat town called Filadelfia appeared like magic in the middle of the wilderness. Here live a community of Mennonites, a Christian fundamentalist sect who arrived in the Chaco a century ago, from Canada and the Soviet Union. Now, as then, they raise cattle and speak their distinctive Germanic language.

Their town museum tells stories of dreadful hardship in the early days. In recent times, however, they have prospered by clearing thorn forest and pioneering the ranching methods now spreading across the Chaco. They produce two-thirds of Paraguay's milk and much of its meat. Along the way, they have swapped pious poverty for air conditioning, four-by-fours and big-box stores full of garden furniture. Besides telling the Mennonite story, the museum was full of stuffed animals of the kind common in the Chaco forests before the cattle arrived. There were armadillos and boa constrictors, skunks and maned wolves, a giant anteater and a six-foot caiman. It was the nearest I got to seeing such animals on my journey. The old, wild Chaco is disappearing.

CAN THIS RAMPANT DESTRUCTION to supply the global commodities trade be halted? The biggest names in the trade say yes. Indeed they promise that the process has been underway for several years. In September 2014, the UN secretary general, Ban Ki-moon, presided over the New York Declaration on Forests. Governments and global corporations from around the world promised to halve what they called "net deforestation" from their territories and supply chains by 2020, and eliminate it entirely by 2030. Many of the biggest commodity companies went further, promising to be deforestation-free by 2020.

Up there at the podium were retail giants like Walmart and Marks & Spencer, food processing moguls such as Kellogg's, Nestlé and Unilever, commodity traders such as Cargill, financiers such as Barclays Bank and pulp giants such as AP&P. Some environmentalists were on the podium too, applauding the good guys. Others dismissed the declaration as corporate greenwashing, with pledges designed primarily to head off legal regulation. They pointed out that many of the worst corporate offenders—including all of the "big five" companies exporting beef from Paraguay—made no zero-deforestation commitments.[8] Still, it seemed there was movement, and maybe the first movers would drag the rest of their industries with them.

A year before he signed the declaration, I had interviewed the boss of Unilever, Paul Polman, at his office beside the Thames River in central London.[9] His company is shelf-stacker-in-chief for the world's supermarkets, producing everything from Ben & Jerry's ice cream to Hellmann's mayonnaise and Lipton tea. Its products consumed more palm oil, probably the biggest cause of rainforest destruction in recent years, than any other company's products. Even so, Polman was playing the green card. He promised a hundred-percent-sustainable supply chain by 2020. "Consumers have sent companies a clear signal that they do not want their purchasing habits to drive deforestation, and companies are responding," he said.

The trouble was that the detail was not there. He could not tell me how his aspirations would be met. So I began my article with the words "Paul Polman is unafraid of making promises he may never be able to keep." So it has proved. Not just for Polman, but for most of the others in New York that day. The truth was they hadn't thought through whether they could deliver. Supply chains are complex. Palm oil is grown not just in big plantations, but by millions of smallholders across the tropics. Unilever was genuine in its ambition, I believe, but really had no idea how it was going to clean up that supply chain when, as one of Polman's top executives told me, "our dirty secret is that we often don't even know who our suppliers are."

By the 2020 deadline, Unilever claimed to have eighty-nine percent of its "forest-related commodities" sustainably sourced. Some other big names, such as Nestlé and Marks & Spencer, were making progress, according to an assessment by Global Canopy, an Oxford-based NGO tracking the world's pledges to zero deforestation.[10] Progress was best among companies with brands to defend and in industries that had been targeted by NGO campaigns. It seemed that reputational risk was at the heart of progress. Certainly the promises made in New York by the anonymous finance houses and hedge fund managers counted for nothing. Indeed, the boardroom voting record of some suggested that they were blocking efforts by the managers of the more progressive consumer brands to clean up their act.[11]

Another initiative of Global Canopy, called Trase, has used shipping bills, customs records, corporate statements and other public data to chart in detail more than 300,000 supply chains taking soy from Brazilian farms to customers across the world. In 2020, it found that big commodity traders such as Cargill, Bunge and ADM, whose ships control seventy percent of the global trade in soy, taking it around the world to feed livestock from Chinese pigs to British chickens, were still prominent buyers from regions where the farmers were engaged in widespread deforestation. Those regions included Mato Grosso, Brazil's biggest soy-exporting state, where Trase found that a fifth of both EU and Chinese soy imports in 2018 came from farms partly made up of land illegally deforested since 2012.[12]

Of course, the corporations never quite said in New York that they would eliminate all deforestation. They said they would end "net deforestation." Meaning they could carry on razing primary forests so long as the losses were recompensed somewhere else by replanting. Still, it was shocking to find in 2020 that during the five years since the New York Declaration the gross annual loss of primary forests worldwide had actually gone up by forty-three percent.

That statistic makes it all the more surprising that, amid the continuing rampant destruction by the commodities trade, something else is stirring in the undergrowth. Something rather more optimistic.

# — 11 —

# TAKING STOCK

## *Phantom Forests and Debunking Forest Demonology*

BACK IN THE 1990S, two young British anthropologists, Melissa Leach and James Fairhead, traveled to Guinea in West Africa to study how local communities in remote rural areas were causing, and coping with, deforestation. Before departing, they had read scientific literature reporting how the land in Kissidougou province, where they were headed, had become almost devoid of trees. Locals had stripped the forests for firewood and to make room for crops. It was a story of demon farmers wrecking the natural landscape.

What they found was rather different. Yes, most of the big old forests were gone. But plenty of trees remained, both on farmland and in woods around settlements. When they looked at aerial photos from the 1950s and compared them with recent images, they found that contemporary photographs showed more forest cover, sometimes twice as much as before.[1] "It was quite a shock," Leach told my *New Scientist* colleague Kate de Selincourt afterwards.

She concluded that the authors of their pre-trip reading material had gone looking for forests and missed the trees. If they had included trees on and around farms, they would have painted a very different picture of the Kissidougou countryside. She also concluded that, despite being widely stigmatized as forest destroyers, most farmers in her study area had been nurturing trees on their land. And they had joined forces with their neighbors to plant entire forests around their villages, as sacred groves, woodlots, orchards and windbreaks. Often, the more people there were in the area, the more trees had been planted.

It seems there is nothing special about Guinea. The findings of Leach and Fairhead, who later married, were published a couple of years after a little-noticed study for the World Bank by Phil O'Keefe of Northumbria University. He had concluded that Africa contained twice as many trees as previously believed, most of them on farms. He found four times more trees in Mozambique than estimated by its own government and twice as many in Nigeria. "There is a surprising amount of wood growing on farms in the arid and semiarid parts of the continent, which is vital to life in these areas but which has been largely missed by past surveys," O'Keefe told me.[2]

Gerald Leach of the Stockholm Environment Institute—who, I later learned, was Melissa's father—agreed. He told me he had found a third of the trees in both Kenya and Uganda were on farms. He called them "invisible trees," because commercial foresters and conservationists both ignore them and because, although they are to be found everywhere, they are not usually in clumps big enough to show up on satellite images.

Such issues abound whenever researchers try to assess the state of the world's forests. Another problem is the permanence of forest clearance. When trees are cut or burned, are they gone for good or will they soon grow back? One analysis of satellite images published in a leading scientific journal as recently as 2018 was headlined: "Congo Basin Forest Loss Dominated by Increasing Smallholder Clearing."[3] It reported that eighty-four percent of "forest disturbance" was due to "small-scale non-mechanized forest clearing for agriculture." What the authors glossed over was that "forest disturbance" is not the same as "forest loss." Much of the smallholder clearing will have been for shifting cultivation, where farmers move on after a year or so and allow the forest to recover. Come back in a decade or so and you will almost certainly find as much forest as before and the apparent deforestation will have turned out to be a mirage.

Some researchers are thankfully trying to measure what is really going on at a local level. One conclusion is that in many rural landscapes, trees are constantly coming and going. The situation is typically "highly dynamic," says Iain McNicol of the University of Edinburgh, after analyzing time-series radar imagery across southern Africa. Both forest loss and forest regrowth are much more widespread than previously supposed. Losses tend to be visible: they happen quickly and mostly near cities and roads, where demand for firewood and charcoal is high. Gains are less noticeable: they happen slowly and in more remote areas, where fewer outsiders go. Overall, he found that at any one time about half of southern Africa was covered by woodlands, and the figure was broadly stable. But the locations of that forested half changed a lot.

Edward Mitchard, also of the University of Edinburgh, goes further. His own field research in forests north of the Congo basin has found widespread increases in tree cover, including in Ivory Coast, Gabon, Cameroon and Ethiopia. He has found extensive "woody encroachment," where forests expand onto savannah grasslands and existing savannah woodlands become woodier.[4] He told me that in some places, "the forest is coming back really fast."

One explanation for this turnaround from deforestation to reforestation is that farming is in decline in many African countries. Growing crops and raising livestock are, as many people there will tell you, "an old man's activity." Young people are moving to the cities or taking industrial jobs. As the older generation dies, farmland is often simply abandoned. Meanwhile, predicted booms in commercial cultivation of oil palms and biofuels in Africa have so far not happened.

THE MORE YOU DELVE into the data on deforestation, the murkier the statistics become. I only discovered the extent of the continued confusion when I compared the results coming out of the two main public sources of data on global deforestation. They have become increasingly contradictory.

One dataset, the Global Forest Watch (GFW), is compiled from satellite images by the World Resources Institute, a Washington think tank.[5] It has long painted a gloomy picture, putting the decline in tree cover in 2017 at 72 million acres, more than double the rate at the turn of the century. The other, the Global Forest Resources Assessment (FRA), is compiled from government inventories by the Rome-based UN Food and Agriculture Organization.[6] It is much more optimistic. It estimates annual losses to be not much above a tenth as great: just eight million acres. It says deforestation rates have declined recently. The drastic difference extends to individual countries. In the United States, China, Australia, Canada, Russia and several other major countries, the FRA shows stable or increasing forests while the GFW shows often substantial losses.[7] What on Earth is going on? Forgive me if things get a bit nerdy here. It is not simple. Nobody is deliberately lying, but what you conclude depends on what you measure.

Nowhere is the statistical and narrative standoff so marked as in Canada, where the difference involves millions of acres. I took a deep dive to try and make some sense of the massively contradictory statistics on offer. One view sees the world's second-largest country dominated by huge tracts of forests that are virtually unchanged since the last ice age. The counterview sees it as a hotbed of deforestation, with one of the highest cutting rates in the world. Both perspectives contain some truth; but both give a misleading impression.

Canada has almost a tenth of the world's forests. Hundreds of billions of conifers make up the boreal forests of the north of the country. They occupy around 750 million acres, in a belt some seven hundred miles wide between the cold, treeless Arctic tundra and warmer lands under the plow to the south. Often the trees are in big blocks, undisturbed by roads, mines, pipelines or other development. The Canadian boreal zone is often called the largest intact forest on Earth, and won Canada the second-highest score, just behind Russia, in a 2020 index of "forest landscape integrity," compiled by the Wildlife Conservation Society.

But if the forests appear intact, they are not untouched. "The romance of the lumberjack still looms large in the Canadian psyche," says Jean-Sébastien Landry, a geographer at McGill University.[8] Each year around one million acres of boreal forest is cut.[9] Some two-thirds of these forests have already been logged at least once. Around a half, including a quarter of the great boreal forests, are currently earmarked for logging, usually by clear-cutting. Several companies hold individual logging rights to areas larger than Switzerland.

The disaster narrative often rests on visual images of the activities of these companies. They are compelling. I have a large-format picture book on my shelves, first published by the Sierra Club in 1993 and still in print, called *Clearcut: The Tragedy of Industrial Forestry.*[10] It has chapters on "Cutting Down Canada" and "Canada: Brazil of the North," accompanied by searing wide-angle photographs of felled forests in British Columbia, Alberta, Saskatchewan, Manitoba, Quebec and elsewhere. In 2020, the Wildlands League, a Canadian nonprofit, unveiled a new set of similar images. "We've been told over and over that Canada has a near-zero deforestation rate," said Janet Sumner of the Wildlands League. But "the logging scars are suppressing the natural renewal of forest."

The logging is real. "Canada is responsible for the greatest total loss, acreage-wise, of primary forest in the world," concluded one 2019 analysis by Frederic Beaudry, an ecologist at Alfred University.[11] He quoted data from GFW that found clear-cutting of around a hundred million acres, or more than ten percent of the total forest, over the previous two decades. British Columbia had lost the most.

But the government says all this talk of forest "loss" is a lie. "Canada's deforestation rate is among the lowest in the world," it says on a website devoted to debunking deforestation "myths." Despite media claims, it says, "harvesting, forest fires and insect infestations do not constitute deforestation... since the affected areas will grow back."[12] Some trees are replanted; some regrow naturally. But in time, they mostly return. Permanent forest losses are less than 0.2 percent per decade, the government says. Moreover, more than ninety percent of Canada's forests are on public land, which must by law be replanted and restored through natural regeneration.

So is this an ecological disaster, or sustainable harvesting? The government is right that Canada is not like the Amazon or Indonesia, where logging is almost a by-product of clearing land to grow palm oil or soy, or raise cattle. Outright large-scale deforestation is rare, at least away from mining areas and regions flooded by large hydroelectric dams. Even the largest clear-cut areas will usually regrow eventually—though the pace is slow in the cold north. GFW, despite its scary stats on logging, agrees that only 0.5 percent of what is cut—mostly lumber for construction, plywood and paper—results in permanent deforestation, with the land adopted for other uses. By most estimates, around ninety percent of those areas of Canada covered by forests after the end of the last ice age remain forested. That is more than a third of the country.

But that does not give the self-styled myth-busters a home run. They obscure as much as they reveal. Forest cover is an extraordinarily incomplete measure of forest health. Environmentalists say that loggers are removing "primary" or "intact" forests, and that what regrows is a crude parody of what was there before. Even if they are recovering all their carbon, they are not recovering their biodiversity. At least not swiftly. Landry agrees. Most of the country's forests are under some form of modern-day management that "has profoundly modified its function and properties," he says.

Another crucial factor is the creeping fragmentation of the forests. Even with only a fraction of a percent cut every year, the roads and other necessary logging infrastructure are eating away at the wildness of the boreal forest regions. So while it is true that around two-thirds of the country's forested area remains for now in blocks greater than 2.5 million acres, the big blocks are gradually being broken up.[13]

That matters a lot, especially for wildlife. Bigger beasts such as caribou and wolverines require large tracts of land to thrive. If the forest is fragmented, they will depart, and even as the forests regrow they may not return, says Beaudry. Instead, alien species will invade along the road networks, "as will hunters, mining prospectors, and second-home developers. Perhaps less tangibly, but just as importantly, the unique character of the vast and wild boreal forest will be diminished."[14]

Up close, most logged forests are far from their former selves. The government suggests that clear-cutting does little more than mimic wildfires, which have always swept through these forests every century or so. But that is not right, say the government's own ecologists, because the natural processes of forest renewal after wildfires are very different from the regrowth after clear-cutting.

Logging removes nutrients from the forest ecosystem, whereas fire recycles nutrients. Nothing departs. Fire also stimulates the reproduction of the many fire-dependent trees, such as lodgepole and jack pine, as well as blueberries and huckleberries that attract wildlife. Logging does not. Forests that regenerate after clear-cutting "retain less biological and structural diversity than those originating from natural disturbance," and so have "reduced biodiversity and resilience," according to a review of the health of the country's boreal forests headed by Sylvie Gauthier of Natural Resources Canada.[15] So this government department appears to disagree with the government's own online myth-busters.

Suzanne Simard, the BC-based ecologist who discovered the subsoil wood-wide webs that sustain most forests, says clear-cutting can dismember these networks. What trees are still able to return will lack these fungal support services. They will be more vulnerable to disease, drought and much else. How long might the networks take to return? Nobody yet knows.

So what is happening in Canada's great forests may not be widespread deforestation, but it is certainly substantial forest degradation.[16] The GFW claims that in recent times Canada has been responsible for more than a fifth of the world's forest degradation, higher than any other nation. It would be hard to disagree.

THE STORY OF CANADA and its remote boreal forests may be particular, but you can see parallels in the confusion over what counts as deforestation that play out in the tropics and elsewhere. So who should we believe? Again, my deep dive found some surprising answers.

On the face of it, you might mistrust the government inventories and believe the satellite data. Surely satellites can't lie? And governments are often economical with the truth. For instance, some classify oil palm plantations, rubber trees and bamboo stands as forests, while others don't. Some call anywhere with more than a ten-percent covering of tree canopy a forest, while others require thirty percent. Depending on your definition, a region of grassland with patches of woodland—such as the Cerrado in Brazil—may or may not be forested.[17] So removing the woodland and planting crops may or may not count as deforestation. The FRA rules, which were drawn up by foresters rather than conservationists, even allow countries to list treeless areas as forests, provided they are earmarked for planting.[18] "Countries can clear massive amounts of forest and still claim that deforestation has not occurred," concludes Peter Minang of the World Agroforestry Centre in Nairobi.[19]

There are other reasons for the difference between the deforestation databases. One factor is that satellites could be just as wayward as human analysts. Satellites can be dumb. They see only what they are programmed to look for. The algorithms rule. If they are written only to identify patches of forest greater than an acre, which is typical, they will miss a lot of trees on farms and along riverbanks.

More problematically, even big forests may go unobserved. When Jean-François Bastin, a geographer now at the Swiss Federal Institute of Technology in Zurich, trawled through high-resolution satellite images not previously examined during global surveys of forest cover, he found 11.5 billion acres of previously uncataloged forests.[20] They were in eleven mostly arid regions, from the southern fringes of the Sahara to central India and Australia. They were enough to boost the current estimate of global forest cover by ten percent. Bastin did not say whether these forests were increasing or decreasing in extent, but they certainly hadn't been taken into account. It seems that the programmers assumed that drylands had no trees, so they never programmed the satellites to look. Similarly, Martin Brandt of the University of Copenhagen analyzed satellite imagery of the western Sahara and found 1.8 billion previously uncounted trees, or 5.5 for every acre.[21]

The satellites may suffer an even bigger blind spot. As Iain McNicol noted, they cannot readily see forest regrowth. Deforestation is sudden, complete, usually concentrated and easy to spot when comparing satellite images from one year to the next. By contrast, forest regrowth—whether natural or the result of planting—is slower, more patchy and dispersed, making it much harder to identify from year to year.[22] Any estimate of the state of the world's forests needs to know if they are regrowing, but we have staggeringly little information about this. Once cleared, is a forest gone for good or will it return?

Philip Curtis of the University of Arkansas dug deeper into the satellite imagery to look for some answers. What he found creates a whole new geography of deforestation. For one thing, only about a quarter of forest loss so far this century is obviously permanent, with the land taken for commercial agriculture, mining, infrastructure or urban expansion. The remaining three-quarters divide equally between forests consumed by wildfires, cleared temporarily for shifting cultivation and logged but with clear evidence of regrowth.[23]

Different regions had different mixes. In North America and Russia, the apparent loss was potentially temporary, mostly due to wildfires or logging. Much the same was true of Africa, where shifting cultivation accounted for more than ninety percent of the trees that disappeared from one image to the next. But in Southeast Asia and Latin America, more than sixty percent of deforestation looked permanent, as forests mostly made way for agricultural expansion. Curtis says his study dispels the misconception that any tree cover loss represents real deforestation. In fact, it suggests that three-quarters of deforested land may be regrowing right now, or could be encouraged to regrow with little economic disruption.

Clearly, the apparently simple data on global deforestation hide a complex story. Equally clearly, neither optimists nor pessimists have a monopoly on the truth. But in 2018, two of the GFW's data compilers seemed to disown some of the more pessimistic takes on their data. Matthew Hansen and Peter Potapov of the University of Maryland concluded in a review of land-use changes over the past four decades that "contrary to the prevailing view that forest area has declined globally," tree cover had increased by 550 million acres, or seven percent.[24] Loss continued in the tropics, but it was now being more than compensated for elsewhere, from tree encroachment onto farmland in Europe to organized tree planting in China, and natural spreading in the far north and some mountain areas as the world warmed.

Forest cover is not the only measure that matters, of course. Quality is as important as quantity. Nobody would say that Siberian conifers compensate ecologically for lost rainforest. Moreover, Curtis concedes that his analysis cannot distinguish between natural regrowth and commercial plantations. Much of the new tree cover will certainly be the latter. According to the FRA, between 1990 and 2015 the area of planted forests worldwide grew by 270 million acres, an area the size of Ethiopia. But Curtis's analysis does suggest a story different from that of unbridled forest loss. It also raises another question: Even where natural recovery is going on, can the new be as good as the old?

This is a hot topic among forest ecologists. For some, regrowth can never replace "intact" forests. Potapov calls intact forests "the last frontiers of wilderness." His widely accepted definition is "a seamless mosaic of forest... with no remotely detectable signs of human activity and a minimal area of fifty thousand hectares [123,000 acres]." This, he figures, is an area "large enough to maintain all native biological diversity," including large carnivores such as bears and tigers, which need wide hunting territories. This is a gold standard for forests, a higher order of quality even than terms such as primary forest or old growth.[25]

Potapov says that in 2018 only around twenty-two percent of the world's remaining forests qualified for the accolade, a sharp drop from almost thirty percent in 2000. At that rate, they could be all gone by 2070. Just under half are in the tropics, with most of the rest in the far north. Two-thirds are in just three countries: Brazil, Russia and Canada. A third are within Indigenous reserves, whereas only twelve percent are in national parks and other government-protected areas.[26]

I wonder if this gold standard is real or useful. After all, the remote Amazon is vast and might seem to be intact. Potapov includes most of it in his mapping. But, as we have seen, this intactness is an illusion. It is full of planted trees, many growing on human-made soils. That sounds to me like a decidedly "detectable sign of human activity." To dismiss the value of much of the Amazon on such grounds—to denigrate the ecological integrity of the most biodiverse forests on the planet—would clearly be perverse. Equally, there could be great ecological value in many other forests that don't meet Potapov's threshold. There is a danger of making the "best" the enemy of the "good." By prizing intact forests too highly, we may unwittingly jeopardize the future of the rest, the great and growing expanses of other forests.

Non-intact forests are often dismissed by the purists as "degraded." There are a lot of different definitions of degraded forest, but whatever the criteria, they too transpire water, shade the planet, store carbon and harbor biodiversity. In many places, species hang on even as most of the forest goes. To take one example: after the 1960s, the small Central American state of El Salvador lost ninety percent of its forests, mostly to coffee and sugar plantations. Much wildlife was lost, of course, yet only three out of its five-hundred-plus bird species disappeared, according to Claude Martin, former director of WWF International.[27] This suggests that even a limited amount of forest can sustain a vast variety of wildlife.

There are other such antidotes to the purist ethos. Researchers in Indonesian Borneo found that patches of forest as small as 250 acres isolated within palm oil plantations provide refuges for endangered species such as sun bears and orangutans. Indeed they are vital to the survival of these species. Three-quarters of the world's orangutans now live not in pristine jungle but on Indonesian timber and palm oil concessions.[28] Half the remaining mountain gorillas are found in the Bwindi National Park in Uganda, which has been repeatedly logged. In fact the gorillas reportedly prefer the more open logged terrain to natural closed forest.[29] "While intact habitats are important," says Karel Mokany of the CSIRO, Australia's national science research agency, "fragmented and degraded habitats can prove essential to conserving biodiversity, because they may support species that have lost much of their habitat elsewhere through anthropogenic modification."[30]

It is routine among conservationists to exclude recently logged forest from consideration for protection. But is this wise? Benjamin Ramage, now at Randolph-Macon College in Virginia, found that logging increases the biodiversity of forests as often as it decreases it.[31] He is not alone in this belief. After reviewing more than a hundred studies of logged forests, Francis Putz of the University of Florida concluded that "85–100 percent of species of animals, birds, invertebrates and plants remain." He decried the tendency to describe such forests as "degraded."

David Edwards, now at the University of Sheffield, took a similar view. In a study of the effects of logging in Sabah, a Malaysian state on the island of Borneo next to Sarawak, he found that more than three-quarters of bird species present in unlogged forest were still there after those forests had been selectively cut for their most valuable species not once but twice.[32] Such forests, he concluded, have a "crucial role in tropical nature conservation" and are "too vast, vulnerable and important to ignore."

Across the world, more than a billion acres of tropical forests, an area larger than India, undergo regular logging of selected trees. Many are subsequently abandoned. In Indonesia, formerly logged forests covering an area larger than Germany have been classified by the government as "degraded." As a result, despite their huge ecological value, this arbitrary distinction means they do not qualify for protection and may face clearance for plantations of oil palm or acacia.

This is a huge and ridiculous own goal by conservationists. But let's look on the bright side. These new findings offer hope too, because they show the ecological vigor of "degraded" forests. They also suggest that if they are left alone they will mostly recover.

And, as Curtis pointed out, most are being left alone. While the world continues to lose the great forests of the tropics, around three-quarters of that loss is being recouped by new trees regrowing naturally on land that has been deforested but not taken over for farming, infrastructure, urbanization or anything else.

That hardly means we are out of the woods. Intact forests clearly have value that cannot quickly be restored, so halting their loss should remain a priority. But, given time, much that has been lost can be regained. That is of huge importance for fighting climate change and protecting biodiversity. It requires, I believe, a global campaign to protect and nurture these "degraded" lands, as the primary means of reforesting the planet. Given the chance, nature will do much of the work. We will explore this in more detail in the remainder of the book.

PART III

# REWILDING

---

Our planet is scarred by huge swathes of deforested and degraded landscape. But everywhere too we can see forests springing back to life, from Costa Rica's ecotourist jungles to the Scottish Highlands; from the sylvan delights of New England to the foothills of the Himalayas. Europe is greener now than it was a thousand years ago. Amid the carnage, the great restoration is underway. Governments have made big promises to plant a trillion trees. But would nature thank us? Or should we step aside and allow nature herself to decide where and when to reseed the world's forests?

---

— 12 —

# FROM "STUMPS AND ASHES"

## *America's Forest Renaissance*

LITTLE MORE THAN a century ago, there were precious few trees in Pennsylvania.[1] It was worse than Mato Grosso today. According to official archives for 1895, stands of old-growth pine and hemlock that had once covered most of the state were down to a few hundred thousand acres, clinging on to rapidly eroding hillsides. Decades of clear-cutting to supply everything from ships' masts to charcoal, and construction materials to tannin, had left one of the founding states of the US arboreally bankrupt. The Appalachian Mountains were covered in "stumps and ashes," as one contemporary put it.

Then the state authorities created a Department of Forestry, began buying land to preserve what trees were left, and established nurseries and trained foresters to begin planting afresh. They continued, right through the New Deal in the 1930s, when job-creation schemes planted fifty million trees, and beyond. Today, the tanneries and most of the sawmills are gone, and almost sixty percent of the state is once again tree covered. There are seventeen million acres of forest—not all of them with native species, but most of them in good health. Red maple and black cherry thrive.

And while the area of forest has now stopped growing, as the new trees age, their size increases. In total, they now weigh above a billion tons.[2]

Further north, much of New England has been even more radically remade. Here nature has done most of the work. Centuries of European settlement and the expulsion of Indigenous peoples had seen forests cleared for timber and pulp, and replaced by farms and pasture. Deforestation never got as bad as in Pennsylvania, but New England's forest cover was below thirty percent by the mid-nineteenth century. Then, as America industrialized, as people moved to the cities, and as food production shifted west, many farms and pastures were abandoned. The Massachusetts timber industry collapsed by two-thirds in the half century to 1950. By that time, trees were growing on Rhode Island six times faster than they were being cut down. Ten million acres of forests were added across the region between 1910 and 1970.

Today most of New England is forested again—seventy-eight percent in Vermont, eighty-four percent in New Hampshire and ninety percent in Maine, where just three percent of the land is still cropped. Connecticut, which was almost entirely tree covered in the seventeenth century, but was reduced to twenty-five-percent forest cover by 1820, is now back up to fifty-five percent. Whole landscapes that were once farmed are now entirely covered in trees. Photographs of Swift River Valley in central Massachusetts in 1890 show nothing but fields and farm buildings. Now there is almost nothing but trees.[3]

The species make-up of the forests is somewhat different, partly because of the suppression of fire and partly due to the arrival of exotic pests. Some old staples such as white pine and red maple have resurged, but beech and hemlock have been much slower to recover. Still, ecologist Jonathan Thompson of the Smithsonian Institution calls it "remarkable that native species predominate and the forest looks in many ways as it has for millennia."[4]

It is true that suburbia has recently begun to encroach. But the overall impact is small so far. New England's forests have been retreating by around sixty acres a day, says David Foster, director of Harvard Forest, a research institute. If that rate continued, the region would not run out of trees for a thousand years. The landscape "has undergone one of the most remarkable histories of transformation worldwide," says Foster.[5] He compares the forests' recovery to the resurgence of the Guatemalan jungle after the collapse of the Maya empire.

Foster thinks the trees are in good hands. Big forest barons, such as Maine's largest landowner, J. D. Irving, remain. But, alongside them, there has been a revival of the tradition of town forests that dates back to the seventeenth century. Now called community forests, they are estimated to number over five hundred scattered across the region. Some date back to the original town charters, but most are much more recent.

The town forest in Gorham, New Hampshire, was planted eighty years ago to protect the community's water supply, but now delivers timber and acts as an outdoor classroom for the town's public schools. The Farm Cove Community Forest, stretching across 27,000 acres of northern Maine, was purchased in 2004 by a land trust to protect public access to the forest and lake shores. Others secure wildlife corridors, maintain bird habitat or support local wood-fuel supplies. So we should not fear the disappearance of the magical sight of New England fall trees changing color.[6]

THERE IS A LOT of mythology about how forested North America was before Europeans showed up. The received wisdom has long supposed that there was endless pristine forest. In the nineteenth century, American poet Henry Longfellow popularized the idea that Europeans stepped onto a continent covered by "forest primeval... undefiled by man's hands."[7] In reality, as we saw in chapter six,

Native Americans managed the land energetically. Much of what we may still imagine to be natural was planted, or managed with fire.

That includes even the ancient sequoia groves of California. The tallest, largest and among the oldest trees in the world "were grown and tended by local people," says California fire researcher Lee Klinger. It is obvious if you look carefully, he says. They cluster in tight circles of trees that grow in soils containing abundant remains of ancient human activity. Native management and encouragement of fire helped ensure that they germinated, since their cones require fire to open and release seeds that grow best in freshly burned soil. The groves were "planted many thousands of years ago and have been carefully tended ever since," according to Klinger. California's Indigenous communities "were not simple hunter-gatherers, but, instead, were sophisticated farmers who practiced sustainable silviculture."[8] The sequoia groves are their greatest legacy.

After Indigenous populations collapsed, the managed landscapes once again became overgrown. Forest cover across North America probably reached a peak in the late seventeenth century in the east and the late eighteenth century elsewhere. Future American president John Adams wrote in his diary in 1756 that the "whole continent was one continued dismal wilderness." As the colonists began to spread west in numbers, they wanted rid of this wilderness, and to create a new model landscape fit for productive agriculture. Agronomists developed theories about how clearing the trees would be good for the land. For a long time it was widely thought that the way to moderate climate, make rain and deliver a more livable environment was to ensure that, as Adams put it, "the forests are removed, the land covered with fields of corn, orchards bending with fruit, and magnificent habitations of rational and civilized people."[9] So that is what they did.

The task was immense. Typically, it took one man and his ax around a month to clear each acre. But acre by acre the forests were cleared. Most of the land was soon tilled. And most of the cut wood was used for construction and as fuel. The latter dominated. In a study for the Forest History Society, Douglas MacCleery, formerly of the US Forest Service, found that in rural America, "in a single year more wood went up the chimney in smoke than had been used to build the house that was being heated."[10] Meanwhile, wood charcoal fueled the thousands of iron-making furnaces that sprang up almost everywhere.

But the peak of timber exploitation passed. Trees came to be seen as either financial, communal or ecological assets to be nurtured. Three of America's top six landowners are foresters: Maine's Irving family, the Emmerson family's Sierra Pacific Industries in California and the Reed family's Green Diamond Resource Company in the Pacific Northwest and South. Collectively they have nearly five million acres, almost the size of New Jersey. Even now, in the days of technology billionaires, cutting trees can still make you seriously rich. Archie Aldis "Red" Emmerson is reportedly worth $4 billion. Famously, he gets richest when the forests burn, and he can send in bulldozers and trucks to salvage charred timber, often from national forests. His teams arrive so fast, he once claimed, that after one California fire "we had trucks coming down the road that had flames on the back."[11] Forest ecologists complain loudly that the charred timber should be left to nourish natural forest recovery, but still this practice goes on.

Beyond the world of timber oligarchs, conservation has created a string of forested national parks, and there is still room for hundreds of community forests and thousands of small family woodlots. Typically just a few tens of acres each, they cumulatively make up thirty-eight percent of the nation's forests and cover an area more than one and a half times the size of Texas.[12]

But the reforesting of America has in many places been driven largely by another factor entirely—declining demand for land to grow crops. The abandonment of farmland started in the northeast as early as 1850, when railroads allowed cheap crops to be imported from the Midwest, and the introduction of motor vehicles reduced the need to grow food for horses and other draft animals. This was accompanied by reduced demand for timber as coal became a cheaper fuel for industrialization. The process continues. In the past thirty years, arable land in the US has declined by almost a fifth, mostly replaced by trees.

With a third of the US now forested, tree growth exceeds tree harvest by an amazing forty percent, says MacCleery. The average volume of wood in American forests is fifty-percent greater than in 1953. In the eastern states, it is as much as a hundred-percent greater.[13] "The entire eastern United States was deforested two hundred years ago," says Karen Holl of the University of California, Santa Cruz. The region now has some ten million acres of forest. "Much of that has come back without actively planting trees."[14]

In the Midwest from Indiana to Kansas, and Iowa to Illinois, trees are recolonizing land once cleared for crops, or they have been planted as part of land restoration projects, often on land previously desecrated for surface mining. Monroe County in southern Indiana has seen an increase in forest cover from thirty-nine percent in 1939 to sixty percent today.[15]

In the South, most states are predominantly covered by forest. The region contains two-thirds of the country's tree plantations.[16] Farms of fast-growing pines prosper, often on former cotton fields or grazing land. (The cattle now live in feedlots and are fed imported grain.) Two-thirds of Mississippi is now forested, five percent up on the 1990s. Much of that is intensely harvested—whether to provide pulp for paper mills or pellets for biomass to fuel power stations. But there are also efforts to bring back the natural swamp forests that once filled the bottomlands down by the lower Mississippi river. Many of these swamp forests were owned by tribes such as the Chickasaw and Choctaw—until they were grabbed and cleared for cotton plantations, and later for rice and soybeans.[17]

According to the US Forest Service, over the past three decades the country's expanding forests have soaked up around eleven percent of the nation's greenhouse-gas emissions.[18] Even in the Pacific Northwest, and despite the wildfires, the forests remain carbon sinks.

None of this should make us complacent about the state of America's forests. Suburban sprawl eats away at the edges and is now thought to be the number one threat to continued reforestation from New England to California. There are other threats. Encouraged by both climate change and increasingly intense harvesting, disease and pests are taking a growing toll on many native tree species, notably chestnuts and elms. And forest composition is changing, as old-growth pine forests are replaced by deciduous trees. But the history of America's forests shows that change is constant. Even when we encounter and revere "old-growth" forest, it usually turns out to be far less pristine than we imagine. The apparently pristine is typically regrowth, either following the collapse of native human populations half a millennium ago, or after the more recent retreat of farmers.

— 13 —

# THE STRANGE REGREENING OF EUROPE

## *Acid Rain to a New Green Deal*

BACK IN THE EARLY 1980S, when I was starting out as an environment journalist, the biggest story was acid rain—the sulfurous fallout from coal-burning power stations that made raindrops as acid as lemon juice. In many parts of Europe, conifer needles were turning yellow and entire forests were dying of what the Germans called *Waldsterben*. West German foresters said half their trees were damaged, a fifth of them seriously. Traveling the country in 1986, I met Bernhard Ulrich of the University of Göttingen, who took me to the Acker ridge in the Harz Mountains of Lower Saxony, where an eco-hotel had built an observation tower so that tourists could look down on the surrounding trees. "Six years ago, the trees were growing so well that they obscured the view," he said. "Today, as you can see, those trees are dead."[1]

Air pollution from industrial plants got even worse as I went east, through forests in the Fichtel Mountains, past a well-fortified American listening post to within spitting distance of the Iron

Curtain border with Czechoslovakia. The air coming across the border had an acrid smell that reminded me of cats' pee. All around, on both sides of the border, the spruce trees were dead or dying. The forest soils had accumulated so much acid that nutrients like magnesium were being leached out, and toxic metals such as aluminum were liberated into tree roots and streams, where fish too died. "This is our moonscape," said Ram Oren of the University of Bayreuth, who was monitoring the trees. It later emerged that over the border Czech foresters removed dead and dying Norway spruce trees from a hundred thousand acres of the Ore Mountains during the 1970s and 1980s, more than a third of the total forest area. In a wide area on the borders between Czechoslovakia, Poland and East Germany, known as the "black triangle" because of its horrendous air pollution, almost no trees would grow.

After the collapse of communism and the closure of its filthy industrial complexes, air pollution was much reduced. Meanwhile, desulfurization technology was added to smokestacks in the West under EU anti-pollution laws. It made rain much less acidic. The soils are taking longer to shrug off the chemical onslaught and many of Europe's forests are yet to recover fully. Nonetheless, even in the black triangle, the Czech government has been replanting Norway spruce, along with other species less susceptible to acidity, such as beech.[2] There has also been substantial natural regeneration.

The acid fallout from communist rule in central Europe may have been bad for the region's forests, but some features of the cold war were a boon. Europe was divided between East and West by high fences, minefields and watchtowers. Along this no-man's-land, a wilderness of forests thrived, complete with wolves, bears and much else. The death zone was a lifeline for nature. The sealed border created an ad hoc chain of nature reserves stretching from the Barents Sea to the Adriatic. Nature did not know that its ponds were mine craters, nor care that its pesticide-free forest glades were protected by barbed wire. When the continent was reunited, conservationists rushed in to preserve what might otherwise have been swiftly lost. It made the rhetoric of a green peace tangible.

Economic changes in the East since the end of the cold war have had both good and bad effects on forests. The bad stuff is undoubtedly bad. Economic liberalization, often combined with corruption, has resulted in an orgy of deforestation in the Carpathian Mountains, home to some of the continent's largest surviving old-growth forests and up to half its brown bears, wolves and lynx. Often, western Europe is the market for the illegal lumber. In Ukraine, two-thirds of the country's logs end up in EU markets. An undercover investigation found that illegally logged Carpathian beech trees have been turned into tens of thousands of chairs sold widely across Europe and the US by IKEA, the world's largest consumer of wood products.[3] In Romania, another NGO investigation found that the state forestry agency had colluded in illegal felling across more than 900,000 acres of its forests since 1990, much of it in national parks. One of Austria's most prestigious timber companies stood accused of processing the wood.[4] Elsewhere, as we have seen in Slovakia, the demands of power stations for wood pellets have contributed to a measurable increase in deforestation rates across Europe.[5]

These bad-news stories may be the staples of environmental campaigning, but they are not the whole story. Out of the spotlight, something else rather more remarkable is happening. Slowly but surely, nature is reforesting the continent.

Following the reunification of East and West Europe, economic dislocation and the withdrawal of socialist support for farming in the East led to an exodus from rural communities, especially from remote or mountainous areas. In the Carpathian Mountains, from Poland and Slovakia to Romania and Ukraine, one estimate is that sixteen percent of farmland was abandoned in the 1990s.[6] So, even as loggers have ripped up old-growth forests in the mountains, new growth has been returning.

Further north, estimates of farm abandonment have ranged from thirteen percent in Belarus to forty-two percent in Latvia.[7] To the east, some thirty percent of the farms in European Russia have been abandoned since 1990. Across the whole country, as many as 250 million acres of former farmland, an area about twice the size of Spain, are now being recolonized by forests, according to a study by Greenpeace.[8] Technically these forests are all illegal, because the law requires farmers to keep their land free of forest. Legal or not, Irina Kurganova of the Russian Academy of Sciences calls the retreat of the plow in modern Russia "the most widespread and abrupt land-use change in the twentieth century in the northern hemisphere."[9]

Much the same has also happened across western Europe. The politics are different, but the demographics are the same. Rural populations are aging and young people are migrating to cities. In most places, as the fields are abandoned, scrub and trees return. This is the main reason why tree cover in the EU has risen in recent years to around forty-three percent. Italy has added 2.5 million acres. In northern Portugal and across the border into western Spain, big agricultural areas have been deserted. The parish of Castro Laboreiro in northern Portugal now has fewer than a thousand mostly elderly inhabitants. Half its 22,000 acres were once grazing land, but now those pastures are virtually all reverting to forest. Ibex and vultures, wolves and wild boars are returning.

In some places, on good soils, agriculture is still expanding. But Francesco Cherubini of the Norwegian University of Science and Technology estimates that over the past thirty years Europe has seen a net loss of fields equivalent to an area larger than Switzerland.[10] Four times more could go by 2040.[11] That would represent a tenth of all current farmland. Most of the abandoned fields are being colonized by trees. The EU's forests already absorb about a tenth of its carbon dioxide emissions. If trends continue, that carbon sink could double over the next half century, calculates Kate Dooley of the Stockholm Environment Institute.[12] If left to grow largely naturally, forest ecosystems would also be restored. The trick for Europe's policymakers will be to find ways of making good use of those new forests, as well as protecting them.

WE SHOULD PERHAPS NOT BE so surprised about the return of Europe's forests. It is not such a new phenomenon. The continent's forest cover bottomed out around 1850. Since then, Europe has put back almost twice as many trees as were lost in the previous century.[13] Despite its dense population, Europe today has around five percent of all the world's trees, which is not bad on around six percent of the land mass.

It is true that most of Europe's trees are, as ever, in Russia. But Sweden has doubled its forest cover since 1900, to more than two-thirds of the country. Spain has tripled its forest area since 1900; France has doubled its forests since 1830; Poland is thirty-five percent up too. Since the 1850s, mainland Denmark's tree cover has grown fivefold, albeit only to eleven percent. Europeans can feel the effects almost every day. By increasing transpiration, the trees are reducing temperatures and increasing cloud cover.[14] Much of western and central Europe is thought to be around one degree Celsius (up to 2°F) cooler in spring and summer than it would be otherwise.[15]

I can hear many environmentalist friends screaming in my ear. Tree cover is only one way of measuring the health of forests. Europe's oldest and least disturbed forests continue to be plundered. Ecologists regard only four percent of its present-day forests as intact. I agree that these old-growth forests, as the reservoirs of much of the continent's biodiversity, need urgent protection. It is also true that much of the new tree cover, replacing the lost old-growth forest, is not remotely as nature intended. Sweden's near blanket tree cover disguises the fact that planted monocultures of spruce and pine now dominate large parts of that country. "Natural forests are so rare than many Swedes have never seen one in their own country," says Plockhugget, a group trying to reinstate broadleaf species such as oak and beech.[16]

Across lowland western Europe, foresters have often replaced the traditional broadleaf trees with coniferous Scots pine and Norway spruce. Today, Europe has 155 million acres more conifer forests and 109 million acres fewer broadleaved forests than it had in 1750.[17] Many of these forests, particularly in central Europe, are currently suffering badly from invasions of bark beetles. The insects bore into spruce trees, laying their eggs beneath the bark, where the emerging larvae feed on the wood, weakening and often killing the tree. Some believe the invasion could be as devastating for the trees of central Europe as acid rain was half a century ago.

The increasing preponderance of conifers also has an impact on Europe's role as a carbon sink. The new conifers hold less carbon than the old deciduous forests. This fact—combined with the worsening state of soils, more frequent wildfires, disease and intensified harvesting—explains a paradox. Despite a ten-percent increase in forest cover, the continent's present-day forests store much less carbon than their counterparts in 1750. Still, whatever the quibbles about their current carbon content, Europe's forest recovery is underway and, with time, many of the "new" forests will mature to become future "old-growth" forests.

Hopefully, this process can be nurtured and accelerated by the political momentum within the EU for what officials in Brussels call their European Green Deal. Frans Timmermans, the EU vice-president in charge of this, says his mission is to "protect and restore the Great European Forest." He produced a plan to grow the continent's forest cover by a further three billion trees by 2030, "with ecological principles favorable to biodiversity."[18] In 2019, the German government announced its own plan to spend half a billion euros in four years. "Every missing tree is a missing comrade-in-arms against climate change," said agriculture minister Julia Klöckner.

What is less clear is the role of natural regeneration in this great endeavor. While Germany has boasted of its planting plans, it also intends to return five percent of its forests to a completely wild state by letting nature do its thing.[19] It has experience of this. Wide areas of Lüneburg Heath in Lower Saxony have been "renatured" since the departure of British soldiers, who had used it for tank training during the cold war. And on the Königsbrücker Heath, north of Dresden, which had been a Russian military training ground until 1992, the government has put more than twelve thousand acres off limits to all humans. The weather has begun to break up the barracks, concrete bunkers and parade grounds; birch, aspen and pine woodlands have colonized the heath. At least one pack of wolves has moved in.

There is a growing desire to let nature rip and see what happens. We cannot be sure how this will turn out. One school of thought holds that nature will restore a continent dominated by dark, closed forests of deciduous trees—the stuff of myths and legends, of Hansel and Gretel and Little Red Riding Hood. Another school says that there was always a lot less dense "high forest"; that landscapes were often dominated by patchworks of grassland, shrubs and woodlands, ruled by large herbivores such as elks and aurochs, the ancestors of modern cattle.

What evidence there is suggests that the patchwork people are right. Pollen analysis indicates that lowland Europe's ancient forests always contained oak and hazel trees, which need a sunny habitat. Europe's premier rewilding experiment appears to be headed in the same direction. When ecologists left Oostvaardersplassen, an abandoned polder in the Netherlands, to regrow without human interference, they expected that forests would soon cover the land. In the event, grazers kept the tree population down and the landscape much more open. A Dutch polder is hardly a typical European landscape, of course, but it is illustrative of the surprises that come from ceding control. Perhaps rather than trying to second-guess nature, we should leave her to it and see what she comes up with.

BILL RITCHIE WAS PROUD to show off his smallholding on the northwest coast of Scotland. Once it was a scrap of sheep pasture. Now he claims it is "possibly one of the most biodiverse spots for hundreds of miles." All he did was clear out the sheep and watch the trees return. It has taken time. "We have been regenerating forest here since 1991," he told me as we toured his rewilded plot overlooking the Atlantic Ocean. "We planted some oak and ash, a few wild cherry trees, but this is ninety-nine-percent natural regeneration. I want to leave the land to do its own thing." And so it has. His trees attract woodland birds, he says, "and because we haven't put up fences, pine martens and badgers can come in. Many people expected we would be overrun with deer, and the trees would not grow, but without grass, the deer don't come."

Ritchie is not alone. His remote corner of Scotland is being transformed by a community-wide effort to restore its trees. Until a quarter of a century ago, the smallholdings were mean and cramped tenancies, known as crofts, that were used by fishers to raise a few sheep and cattle. If they had any trees, they were usually hazel, grown to make fish traps. The crofts were surrounded by thousands of acres of bare hillsides, covered in heather and rough grass, run by absentee landlords for sheep grazing and deer stalking. The landlords hereabouts were the Vestey family, owners of the international beef trading company of the same name, which a century ago pioneered turning South American rainforests into cattle ranches to supply its chain of butchers.

In 1993, Ritchie and others bought out some twenty thousand acres of Vestey land around their crofts and began creating their own greener and much more biologically diverse North Assynt Estate, north of the fishing village of Lochinver. Some, like Ritchie, are rewilding their own small plots. Collectively, they have planted over 2,500 acres of native woodland and rehabilitated much more, transforming the former rough pasture around their thirteen hamlets, resplendent with Gaelic names such as Clachtoll, Stoer and Balchladich.

The purchase proved a watershed for the restoration of forests in the Scottish Highlands. Since then, many other communities have raised funds to buy out their landlords. Not everyone who takes possession of the land has the same idea about how to use it. Some would like to rewild the land by getting rid of the free-roaming deer that feast on tree saplings; others are more interested in making money from well-heeled deer stalkers. Mostly, however, the two impulses coexist, and all agree that forests provide more jobs than open moorland, whether in vistas for tourists, hazel for fish traps, refuges for stags and cattle, or inspiration for artists.[20]

Ritchie's partner, environmentalist and author Mandy Haggith, drove me into Lochinver to meet members of the Culag Community Woodland Trust, a local charity that bought another piece of the Vestey estate. They spend their weekends planting it with native birch and other trees raised in their own nursery beside Loch Assynt. The wood was growing so dense that, as one activist complained, "I can't ride my pony through there anymore." Meanwhile another local organization, the Assynt Foundation, had bought a much bigger slice of Vestey landholding. It includes four striking mountains and the Vesteys' grand old Glencanisp Lodge. It featured in a 2017 movie starring Sheila Hancock as the elderly *Edie*, who set out to climb one of the sandstone summits, the foundation's CEO, Gordon Robertson, told me. He planned to expand the existing birch, rowan and willow woods, as well as bring in tourists, build artists' studios and establish new croft tenancies for locals.

Taken together, these projects in a remote corner of northwest Scotland make up one of the largest landscape restorations in Europe. (They are certainly a big change from the last attempt to reforest the Highlands, when the UK government a century ago created the Forestry Commission. The idea was to make the country self-sufficient in wood, but the emphasis on maximizing production, and disdain for the old landscape, left a legacy of endless and much-hated conifer plantations.) The projects combine two powerful community-led ideas: to take the land back from big landowners and to reforest it. What began in the crofts around Loch Assynt has turned into a nationwide movement, aimed at creating cultural, environmental and economic assets under the control of local communities. "Restoring the land and the people" is its slogan.[21]

Not every such project in Scotland is a result of collective action. Furniture-company heir Paul Lister has planted a quarter of a million native pines in his 22,000-acre glen north of Inverness. He would love one day to reintroduce wolves to protect his trees from deer. Then he could rip up the fences, he told me when I visited. Anders Povlsen—the largest shareholder in the ASOS fashion chain, and one of several rich Europeans with estates in the Scottish Highlands—is likewise greening his 44,000-acre Glenfeshie estate in the Cairngorms, but is culling deer to allow the forest's return.

Most of the time, however, it is communities and activists rather than lairds who are restoring and nurturing the forests. In the south of Scotland, the Borders Forest Trust is building a small forest empire in the once-treeless Moffat Hills. The activists here are not crofters, but mostly retired academics and conservationists. They bring binoculars as well as spades. By planting native trees from local seeds, they set out to bring back "the rich diversity of native species that existed there six thousand years ago," says trustee Hugh Chalmers. Seventeen years and 600,000 plantings on, the trees are above head height, undergrowth is appearing and a profusion of wild flowers have invaded the gravel beds down by the Carrifran River.[22]

For environmental activists, such projects offer a template for democratic forest restoration and ecological rewilding in many other post-industrial societies in Europe and perhaps beyond. Back at Ritchie's croft, Haggith is not content with seeing the return of pine martens and badgers amid the trees. "What we really need is bears," she said. Bears were once lords of the old forested highlands. They have not been seen for two millennia, but "they are the best seed-spreaders in woodlands," she said. "We need wolves to eat the deer, and bears to spread the trees." That would certainly be rewilding.

# — 14 —

# FOREST TRANSITION

## *How More and More Nations Are Restoring Their Forests*

AMERICA HAS A LOT of land. There is plenty of room for trees. But if Europe, which is more densely populated, can restore its forest cover, how about China? Is there room in the Middle Kingdom for 1.4 billion people and forests too? It seems so. We saw earlier how, after 1998, China banned logging in its natural forests, but there was also a second strand to its policy: a national program to plant trees. Since then, it has recruited more than a hundred million farmers to the task, and claims to have restored around 75 million acres of forest, under the "conversion of cropland to forest program," or Grain for Green.

I went to see one of its biggest and most concentrated efforts, on the Loess Plateau beside the middle reaches of the Yellow River. The plateau is an area of loose sand hundreds of feet thick, as big as France and with almost as many people. Erosion has long been rampant on the plateau. It's not much of a plateau anymore, just a maze of hills and gullies. The river downstream contains more sandy silt than any other river in the world. It clogs reservoirs and causes floods. So Chinese engineers have set to work to halt the erosion and make the Yellow River run clear, by turning the plateau's steep and crumbly slopes into staircases of terraces, on which they have planted millions of trees.

I came on the work unawares. Cresting the brow of a hill in Dingxi county, I suddenly saw terraced hillsides stretching far into the distance. Gangs of men and women were passing saplings by hand up the hillsides and planting them in the hot sun. My guides from the Yellow River Conservancy, a government agency with the task of protecting the river, told me that upwards of thirty thousand workers were engaged in the project at any one time. It was breathtaking, looking down on thousands—probably millions—of cypress trees in serried rows on hundreds of terraces climbing the hillsides all around. Local water chief Cheng Hangjun pointed to barren hills in the far distance. "That is what these hills looked like until last year," he said. "Our work is not yet complete."

For a while, the World Bank funded the Loess Plateau Watershed Rehabilitation Project.[1] Its then president, James Wolfensohn, took a personal interest. Speaking years later in Shanghai, he remembered that, when he first inspected the plateau, "the hills in an area the size of Switzerland had looked arid, terrible, stark with no trees." Now "they've got grass and trees and animals. It's an area where you'd like to have a holiday. It looks like Switzerland." That was a stretch, but I knew what he meant.

How successful has all that labor been? The Chinese government claims a sixty-to-eighty-percent reduction in erosion on the plateau, with more than 300 million tons less sediment reaching the Yellow River every year. That was the plan, for sure. But postmortems are ambiguous about the overall environmental benefits. The trees may have reduced water erosion, but by increasing transpiration into the air, they almost halved river runoff, reducing local water availability.[2] One report written by the Chinese Academy of Sciences at the conclusion of the project warned that "the newly restored ecosystem is weak and relatively unstable. It may revert to its previous state, unless care is taken."[3] Part of the reason is that the ecosystem wasn't "restored" at all. It was made from scratch, by planting trees where there were none before. The lesson is that just planting trees is not an environmental fix-all. Benefits may come with trade-offs. And, yes, the Yellow River is still yellow.

Grain for Green has been credited with increasing China's tree cover by a third and reducing soil erosion by almost half.[4] But its regimented approach, planting the same trees on different terrain and corralling local labor, with little scientific analysis beforehand or care and appraisal afterwards, has been a recipe for poor returns. Especially if the social consequences are not properly addressed. In southwestern China, Grain for Green apparently increased the region's tree cover by thirty-two percent. However, the plantations were almost all on former fields, says Fangyuan Hua, an ecologist at the University of Cambridge. The displaced farmers found replacement fields in a rather predictable way—by clearing nearby natural forests. "The region's apparent forest recovery has effectively displaced native forests," she says.[5] It ended up with more plantations, but fewer natural forests.

Another centerpiece of China's efforts at reforestation has been the Great Green Wall, planted to halt the advancing Gobi and Taklamakan deserts in the north of the country. These expanding deserts are Asia's biggest dust bowls, whipped up into clouds of sand that have regularly reduced visibility in Beijing to close to zero. So the plan is to plant some hundred billion trees in a three-thousand-mile belt from Xinjiang in the west, through Inner Mongolia to the Yellow Sea. The wall is probably the largest ecological engineering project on the planet. It is also one of the few big policy initiatives in China to endure since the revolution in 1949. Initiated by Chairman Mao, and begun as Deng Xiaoping took control in 1978, it rolls on through the era of Xi. It may not be completed till the middle of the twenty-first century.

Tens of billions of dollars have been spent planting species such as saxaul, a native of the steppes of central Asia. With willing laborers in short supply as young rural people move to the cities, organizers have brought in the military to keep work on track. In 2018, some sixty thousand soldiers from the People's Liberation Army were seconded from posts on the border with Russia to plant trees in Hebei province, which surrounds Beijing.

The government says that, thanks to the planting so far, the deserts are now retreating. "Vegetation has improved and dust storms have decreased significantly," Minghong Tan of the Institute of Geographic Sciences and Natural Resources Research in Beijing reported in 2015. "In north China as a whole, we think the environment is getting well."[6] Western geographers are skeptical about such claims, as have been many within China. Jiang Gaoming of the Chinese Academy of Sciences' Institute of Botany called it a "fairy tale."[7] Shixiong Cao of Beijing Forestry University said that as many as eighty-five percent of plantings died.[8] "China's huge investment to increase forest cover seems likely to exacerbate environmental degradation in environmentally fragile areas," he said, because "it has, in many cases, led to the deterioration of soil ecosystems and decreased vegetation cover, and has exacerbated water shortages."

David Shankman of the University of Alabama says the trees are often planted in places that are too dry for them to grow naturally. Some have survived thanks to plenty of initial watering but will not reproduce. Others have sucked up so much underground water that the water table drops, killing local natural vegetation that is more resistant to drought and more effective at erosion control than the introduced trees. So rather than holding back the deserts, the Great Green Wall could accelerate their advance. When I put this criticism to Tan, whose 2015 paper had boasted of the wall's success, I was surprised to find he agreed. The authorities should not focus just on increasing forests, he said. "Grass may be better in most places in north China."[9]

China nonetheless sees itself as a model for regreening the world. It has promoted tree planting as part of its giant Belt and Road Initiative to build infrastructure in developing countries. It has helped Pakistan set up thousands of nurseries for planting what it calls a "billion-tree tsunami" on the border with Afghanistan, with plans to extend it to Kazakhstan, Iran and even Turkey. It has also been taking the lead in UN efforts to push nature-based solutions to climate change. But, while the ambition is laudable, its claims of domestic success are far from clear, and the risks of replicating failure are high. Hong Jiang of the University of Hawaii, a cultural geographer interested in how her native China sees the environment, told me that China's "aggressive attitude towards nature," planting trees where they do not grow naturally, will not ultimately work. "Instead of controlling nature, we need to follow nature," she said.[10]

⁂

THE ARGUMENT FOR GROWING FORESTS continues to energize people around the world, whether the aim is to hold back deserts, prevent floods, green lives, capture carbon, maintain biodiversity, forestall dangerous changes to local climate or simply let nature do her thing. And, where there is a will, it can happen in the most unlikely circumstances.

Take the small Central American nation of Costa Rica. For much of the twentieth century it was a hotspot of deforestation. Its lush rainforests were ripped up to make way for cattle pasture, much of it to feed the voracious demands of American burger chains for beef. At one point, the country reputedly sold sixty percent of its beef to Burger King. Its ranchers were offered soft loans to remove trees and keep up with the demand for meat. I remember reporting that the country had one of the fastest deforestation rates in the world. Tree cover declined from seventy-five percent in 1940 to twenty percent in the late 1980s. The environmental damage was everywhere. Denuded hillsides were suffering landslides, while down in the valleys there were ever-worsening floods.

Then two things happened: beef prices fell and Rafael Calderón, president between 1990 and 1994, got green fingers. After the Earth Summit in Rio de Janeiro in 1992, he called for a "new ecological order" and asked the charismatic Canadian chairman of the Rio event, Maurice Strong, to establish an Earth Council in Costa Rica. Then he got his environment minister, René Castro Salazar, to set up a system for paying landowners to protect surviving forests and nurture new ones that readily grew on pasture from which cattle were removed. Started in 1996 and funded by a fuel tax, that system has persisted through six governments elected on very different political platforms.

Calderón sold his plan as a project to reduce floods ravaging the country and to encourage ecotourism. It worked. Forests now once again cover more than half the country.[11] Some are planted; most are a result of natural regeneration. Floods are reduced and ecotourism is now worth $2 billion per year to the country. More enterprising farmers are building lodges in the new rainforests for visitors who are mostly unaware that the forests they admire were until recently cattle ranches.

Of course, there are caveats. Deforestation hasn't entirely stopped. In Coto Brus, a canton on the country's eastern border with Panama, half of the regrowth was cleared again within twenty years, according to Leighton Reid, now at Virginia Tech.[12] Cattle ranchers have returned in places and some farmers are clearing forest to grow pineapples. Nobody said it would be easy, or that nothing could go wrong. The trend towards restoration is clear, though, and has continued now for a quarter of a century. It can be done. Costa Rica has justifiably become a poster child for how tropical nations can reverse deforestation by letting the forests come back.

Costa Rica is far from alone. Elsewhere, similar changes are occurring. In fact, it happens so often that you could even say it is now standard for countries rising up the scale of economic development to stop deforesting and start reforesting. Environment economists call this journey the "forest transition." When Pekka Kauppi of the University of Helsinki did a simple correlation between national statistics on changes in forest cover and economic status, he found the link was unmissable. Between 1990 and 2015, high-income countries showed an average annual increase in forest cover of 1.3 percent, while middle-income countries managed an increase of 0.5 percent. However, low-to-middle-income countries lost 0.3 percent of their forests each year, and the poorest lost 0.7 percent.[13]

This rule applies to most nations. South Korea almost doubled its forest cover as its economy boomed after the end of the Korean War. In the half century from 1953—thanks primarily to a government-led program begun in the 1970s—its forest cover rose from thirty-five to sixty-four percent.[14] The only wealthy country Kauppi found to be still losing forests was the small oil-rich state of Brunei, on the tropical island of Borneo.

Among developing countries, thirteen have begun forest expansion since 1990, including Chile, China, Costa Rica, El Salvador, India, Thailand and Vietnam. In Panama, the government ran a program for the "conquest of the jungle" as recently as the 1970s, but did a U-turn following concern for the reliability of water supplies needed to keep the Panama Canal full.[15] A twenty-year program, much of it with native species, is now paying off, with an estimated 1.5 million more trees. Nepal has seen a rise in national forest cover of around a fifth in the past four decades.

Interestingly, forest expansion happens even in countries with fast-growing populations. China transitioned from net forest loss to net forest gain soon after 1970, before the introduction of the one-child policy and at a time when its population was growing by a fifth per decade. India has more than doubled its population since 1970, while making the same shift in its forest cover.

We may not always like how the reforesting is being achieved. Chile has spent large sums subsidizing plantations while continuing to lose biodiverse native forests.[16] The Indian government allows developers to fell natural forests so long as they replant an equal area of trees somewhere else. This usually means that biodiverse forests used by poor communities—or forests in watersheds critical for river flow or transpiration, such as the Western Ghats—are replaced by commercial forests.[17] Nonetheless, one way or another, India is finding room for more trees. Its forest plan promises to raise forest cover from twenty-four to thirty-three percent.

So is the best route to reforesting the world just to get rich? Well, maybe. But any environmental halo over rich nations would be misplaced. In general, the well-off have the money to outsource their deforesting as they buy food and other commodities grown by clearing land in other countries. As we have seen, they don't stop deforesting; they just do it somewhere far from home. Europe began regaining its forests in the mid-nineteenth century, just as it started to clear forests in the tropics to grow what it needed. Its consumption of everything from beef and palm oil to rubber and sugar means it remains today responsible for a tenth of global deforestation, with thirty-six percent of that from internationally traded commodities. Between 1990 and 2008, that totaled an area of lost forests outside the EU the size of Portugal.[18]

A global restoration of the world's forests will require rich nations to take responsibility for the impact of their consumption, to both protect existing forests and make room for more.

— 15 —

# TO PLANT OR NOT TO PLANT

## *When Trees Become Part of the Problem*

IN 2019, TREE PLANTING was a global craze. For a while, my email in-box was awash with PR about private tree-planting projects across the world. Ecosia, the world's largest nonprofit search engine, announced it was using its spare cash to plant up to a billion trees across nine thousand sites through the tropics. A bunch of YouTube stars raised $6 million for the nonprofit US-based Arbor Day Foundation. The foundation was set up way back in 1972 and claims to have planted 350 million trees over the past half century, on American campuses, in the acid-rain-damaged Ore Mountains of central Europe, in the rainforests of Madagascar and many other places. But it said it had never seen such business.

Some appeals came with great backstories. Clare Dubois set up TreeSisters after a tree prevented her and her car from tumbling into a ravine in the English countryside. TreeSisters is now busy in Madagascar, India, Kenya, Nepal, Brazil and Cameroon. From Kenya, I heard that the late Nobel laureate Wangari Maathai's

Green Belt Movement had passed fifty million trees, all planted by women in their local neighborhoods. Taking a different approach, Worldview Impact was using drones to scatter tree seeds across Myanmar.

Many such projects don't come to fruition. Back in 2011, somebody called Rodrigo contacted me from a just-launched outfit called Amazing Forest, with a plan for "rebuilding the Amazon rainforest, one tree at a time."[1] With perhaps a hundred billion trees lost from the Amazon, that sounded like a tall order. "We sell trees," the press release said, "to be planted on your behalf, in your name in a designated specific area of the Amazon rainforest... What's cool about it, people get their own tree, which will never be cut down, individualised by a name tag."

They had some map coordinates to a site thirty miles outside Boa Vista in Roraima, one of the least deforested states of the Amazon basin. At sixty dollars per tree, it was a lot more expensive than most tree-planting projects I have come across. Yet when I checked it out eight years later, I found Amazing Forest.com was no more. Indeed I could find no sign of its existence after the launch date. Rodrigo and his partner Ben's other companies, listed as bona fides at the launch, also seem to have disappeared without trace. I'd love to know what happened to the trees—and the money.

I don't want to be too cynical. Probably their aims were laudable. We all know we take a bit of a gamble when we give money to small charities, but you would hope governments were a better bet. Yet now there are growing concerns that government promises to heal the planet by planting trees may sometimes prove almost as ephemeral. Back in 2011, when Amazing Forest was pitching for my cash, most of the world's governments signed up to the Bonn Challenge. It was an initiative of the German government and the International Union for the Conservation of Nature to revive the world's lost forests—to "bring 150 million hectares [370 million acres] of the world's deforested and degraded land into restoration by 2020, and 350 million hectares [860 million acres] by 2030." That is an area a bit bigger than India. Besides planting more trees, the aim was to "restore ecological integrity," "improve human well-being" and soak up as much as two billion tons of carbon dioxide annually.[2]

Initially, progress seemed good. Some forty countries had made formal pledges, covering 420 million acres. Brazil, China, the Democratic Republic of the Congo, Ethiopia, India, Indonesia, Mexico, Nigeria and Vietnam had all promised more than 25 million acres of extra forest. My colleague John Vidal at the *Guardian* called it "a eureka moment" and a "great cultural shift" in which forests "have changed from being dark and fearsome places to semi-sacred and untouchable... From Costa Rica to Nepal, tree planting has become a political, economic and ecological cause, and a universal symbol of faith in the future."[3] Well, up to a point. Many of these countries were increasing their tree cover, but an independent report funded by the German Bundestag concluded that by late 2019 trees were present on fewer than 67 million acres—less than a fifth of the Bonn promise for that date.[4] Other analysts, meanwhile, had some hard questions about the promised ecological integrity, human well-being and carbon capture.

When Simon Lewis of the University of Leeds examined the small print of Bonn pledges made up to mid-2019, he found that forty-five percent of the promised new forests were plantations, mostly of fast-growing trees destined for harvesting in double-quick time to make pulp for paper.[5] Such monocultures made up eighty-two percent of promised planting in Brazil and ninety-nine percent in China. These were not natural forests. They would contribute little to ecological integrity or human well-being, and probably not much to climate either. "Policymakers are... misleading the public," he said. The implication that industrial plantations were the same as forest restoration was a "scandal."

It was a scandal, moreover, that "put climate targets at risk."[6] Why? Because the plantations will ultimately store much less carbon than natural forests. Over the short run, they may outperform the carbon take-up from natural regrowth. But, since they are likely to be harvested within a decade or so, that won't last. And it is the long run that matters in fighting climate change. Through their life cycle of growth and cutting, plantations may on average hold little more carbon than the land cleared to plant them, said Lewis. Restoring natural forests over the promised 860 million acres would remove around 45 billion tons of carbon from the air by 2100, but installing commercial monocultures would remove only one billion tons.

Not everyone is so opposed to plantations. WWF and the World Business Council for Sustainable Development have set up a project called New Generation Plantations, which aims to encourage more environmentally friendly versions. "Plantations can contribute to a sustainable bio-economy," says one of its advisers, Jaboury Ghazoul of the University of Edinburgh. He foresees wood replacing high-carbon materials such as concrete, steel, plastics and cotton, and making feedstock for biofuels and the chemicals industry. "The economy of the future will grow on trees," he said. Maybe so. But such tree farms will often not do the jobs suggested in the Bonn Challenge. Indeed, they may instead take the land that could otherwise be devoted to restoring natural forests, improving livelihoods and capturing carbon for the long term. So while the Bonn Challenge is a great idea for encouraging countries to reforest their lands, as many already are, it does run the serious risk of shutting off initiatives for the kinds of forests we really need.

WE NEED TO LOOK in a bit more detail at this whole issue of tree planting and carbon. Trees have long been promoted as part of the fix for the climate crisis, because they draw carbon dioxide out of the air as they grow. I have on my shelves a report from the US Department of Energy dated 1988 entitled "The Prospect of Solving the $CO_2$ Problem Through Global Reforestation." The author, Gregg Marland, then of the US government's Oak Ridge National Laboratory in Tennessee, concluded that "although looking to forests to solve the $CO_2$ problem is unrealistic, reforestation could indeed play a significant role as one component among a variety of measures taken to address increasing $CO_2$."[7] Thirty-two years on, Marland's judgment holds true.

Trees as carbon sinks now have center stage, following the 2015 Paris Agreement on climate change. Paris promised to limit global warming to well below two degrees Celsius—and preferably 1.5 degrees (or below 3.6 and 2.7°F). The main task must be to end the world's addiction to getting our energy for electricity, industry, transport and the rest by burning fossilized carbon and filling the air with carbon dioxide. Nonetheless, scientists at the UN's Intergovernmental Panel on Climate Change (IPCC) say that the target will be almost impossible without a big contribution from forests. First off, that means ending tropical deforestation, which still produces anywhere between ten and fifteen percent of our emissions. The IPCC says that we will also have to find ways of removing carbon dioxide directly from the atmosphere. Getting trees to photosynthesize more carbon dioxide from the air seems the most feasible option for achieving these "negative emissions."[8] That means more trees.

This is exciting for environmentalists. Nature rides to the rescue. A much-cited paper by Bronson Griscom of The Nature Conservancy and others claims that such "nature-based solutions" could absorb up to twenty-six billion tons of carbon dioxide a year, a third of current targets for curbing the buildup of the gas during the 2020s, though less thereafter.[9] Banishing fossil fuel emissions remains essential, but more trees could make all the difference, as warming nudges up to 1.5 degrees and beyond.

With proper attention to the recovery of natural forests, achieving the Bonn Challenge could make a major contribution to curbing warming through nature-based solutions. One outstanding question has been whether enough land is actually available for trees, whether planted or naturally regrowing. Enter Thomas Crowther. A British ecologist in his mid-thirties, Crowther likes big data and has an eye for a stat that could change a debate. In 2015, he came up with the breezy assessment that the world's forests contained three trillion trees—six times more than previously guessed.[10] Partly on the back of that, he got his own lab at the prestigious Swiss Federal Institute of Technology in Zurich and a funding stream from a rich Dutch NGO.

Four years on, with colleague Jean-François Bastin, he grabbed the world's attention with detailed mapping analysis showing that the world had room for 1.2 trillion more trees on 2.2 billion acres of formerly forested land not currently in use for agriculture or human settlements.[11] Those trees, the pair said, could soak up around 200 billion tons of carbon, right in line with what the IPCC said might be needed by 2050 to limit global warming to 1.5 degrees. It would be cheap too, at around forty US cents per ton of carbon. A handful of billionaires could pay for it, he suggested.

Published in the top journal *Science*, this combination of mapping and number-crunching made both the front page of the *Guardian* and a line in Donald Trump's next presentation at Davos—a rare combination, to be sure. In fact, tree planting was the talk of the World Economic Forum at the Swiss resort of Davos in 2020. Months later, Trump followed up with an executive order promising close to a billion new trees on 2.8 million acres across the United States, albeit mostly planted by corporations and NGOs rather than the federal government. And he didn't see the contradiction between announcing that and, just a few weeks later, exempting the Tongass National Forest in Alaska from a ban on road building in national forests, thereby opening up millions of acres to industrial logging.[12] Still, there is a lot of potential for more trees. The Nature Conservancy estimates that there are twelve million acres of federal lands in need of reforestation, including 7.7 million acres within national forests.[13]

Crowther enjoyed all the attention. As another top journal, *Nature*, put it in a profile a few weeks later: "Publicity, he believes, will get him closer to his goal, and his goal is nothing less than restoring the planet."[14] But a lot of Crowther's peers hated his paper and the publicity around it. They may have been jealous; they certainly found the new kid on the block a bit pushy. In a *New York Times* op-ed, Erle Ellis of the University of Maryland called his contribution "a dangerous diversion [that] takes attention away from those responsible for the carbon emissions," to wit the fossil fuel industry.[15] Crowther's university, despite having organized much of the publicity for him, turned hostile too. It banned him from talking to the media for several months.

*Science* eventually ran several pages of critical responses. A lot of the criticism seemed like sniping and ganging up. Besides the "dangerous diversion" charge, many were nitpicking about the mapping, or argued that Crowther did not investigate other factors necessary for planting trees, such as whether there was enough water, or consider the potential climate changes that planting forests might induce in some places. Such denunciations amounted to little more than that he should have done different research, though other criticism had more substance. To get his numbers up, he seemed to have too readily assumed that grasslands with some trees in them were degraded forests ripe for "restoration," rather than ecosystems in their own right. The loudest complaints here came from Africa, the grassiest continent.

Just before the media clampdown, I interviewed Crowther at his lab in Zurich. The pushback had startled him. "But we only ever had a simple aim, to do what nobody had done before and map all the land potentially available for restoring forests. We never said whether we should or should not plant in specific places—only whether we could. So it is the start of a discussion, not the end." He had another great stat. Restoring forests could capture the equivalent of "two-thirds of all the emissions added so far to the atmosphere by human activity."

After the interview we had a long correspondence about what I could print without getting him fired—which, journalistic independence or not, I didn't want to happen. I thought he had made a great—and, as it turned out, rather brave—contribution to climate policymaking. I agreed with Robin Chazdon, a forest ecologist at the University of Connecticut, that "we need the Crowthers who are out there pushing the limits and giving us big numbers and big potential that will excite people."[16]

What really concerned Crowther's critics was that his framing would unleash a tidal wave of dubious forest plantation projects making big claims for halting climate change through "carbon offsetting." Projects that, as Lewis points out, may have few real benefits in terms of biodiversity and store much less carbon than natural forests. Their main purpose, some suggested, would be to "greenwash" polluters such as oil companies. Companies like European oil giant Royal Dutch Shell, which just a few weeks before Crowther's paper appeared had announced that it was going to spend $300 million planting trees in Spain and the Netherlands to offset emissions from burning its products. And Italian oil major Eni, which said it planned to plant up an area of Africa half the size of England.

To explain, offsetters set up projects that either plant trees to soak up carbon dioxide or keep the gas out of the air by preventing deforestation. They get scientific certifiers to adjudge how much carbon dioxide is kept out of the atmosphere by these projects and sell this benefit in the form of carbon credits to polluters, who can use them to avoid obligations to cut their own emissions. There is a growing global market in these credits, which companies generally buy if they come cheaper than complying directly with emissions limits. So a lot hangs on the credibility of these carbon IOUs.

Some big questions arise. First, when trees are planted, how permanent will they be? Will they be phantom forests that never grow or are swiftly deforested? To be useful they need to absorb carbon dioxide for many decades, rather than succumbing to bulldozers, chainsaws or fires. Second, how can we know whether the new trees have been planted anyway, or would have grown naturally on that land? Third, when credits are allocated for protecting trees, how can we know if they would otherwise genuinely have been cut down? Sometimes the evidence that the credits make the difference is surprisingly scant. A 2015 study of over four hundred forest offset projects found that a quarter of carbon credits were issued for areas that were, in part at least, already in protected areas.[17] So where is the extra benefit? The danger is that business-as-usual conservation ends up allowing business-as-usual pollution. Even if the protection of a particular forest is real—that it would genuinely have been cut down without the extra cash from credits funding the protection—there is still another question. Won't the loggers or farmers who have been thwarted just move on and chop down some other trees instead? Again, if that happens the atmosphere will not gain. Planting trees becomes part of the problem.

Finally, says Nathalie Seddon of the University of Oxford, there is the worry that reforesting for climate mitigation will be handy PR for business-as-usual forestry, giving a green patina to projects that are "coming at the cost of carbon-rich and biodiverse native ecosystems and local resource rights."[18]

ALL THIS PRESUPPOSES that trees are good for the climate because they soak up carbon dioxide, recycle moisture and so on. Generally that is true. But not always. As we have seen, in the great northern forests of the Arctic, the warming effect of the dark canopy exceeds the cooling effect of carbon take-up. There are other places where the same may hold true: deserts, for instance. Dense stands of trees rarely grow naturally in deserts, so planters need to be wary of doing more harm than good.

Two hours' drive out of Tel Aviv, on the southern slopes of Mount Hebron, the Yatir Forest is Israel's largest planted woodland.

It has four million trees, spread across an area a bit smaller than Tel Aviv. The dense stands of Aleppo pine are a substantial dark green smudge next to the dull yellow sands of the Negev Desert. Through breaks in the trees, there are views east towards the Dead Sea, north onto the Palestinian West Bank and south across the Negev, towards the town of Be'er Sheva. Beyond some Bedouin grazing their sheep, I saw a solar power plant glinting through the haze. Somewhere out there too was Israel's secret nuclear reactor and weapons arsenal at Dimona.

For some, this forest is a model for greening the world's deserts. It could be a foretaste of a global effort to halt climate change by planting trees from the Sahara to northern Australia and from the Botswanan Kalahari to the Chilean Atacama. Israeli scientists say such forests could make the deserts wetter, so natural forest might grow without human help. Is this the ultimate trillion-tree dream? Or a desert mirage? I had come to find out.

Trees matter a lot in arid Israel, culturally as well as ecologically. Many Israelis and the Jewish Diaspora around the world celebrate the annual "holiday of trees," Tu BiShvat. Creating forests is, as one scientist told me, "a way of saying we are here"—of making the desert forever a part of the state of Israel. Tree planting is a popular way to honor a loved one. "It's practically a Zionist commandment," Jay Shofet of the Society for the Protection of Nature in Israel (SPNI), the country's oldest conservation group, told me in his office in Tel Aviv. "The image of a forested Israel has always fired the imagination of well-intentioned Zionists."

The planting of the Yatir Forest began in 1964, as part of a wider effort to green and occupy an area of Israel bordering the West Bank. It was undertaken by the Keren Kayemeth LeIsrael–Jewish National Fund (KKL–JNF), a nonprofit group created as long ago as 1901 to buy land for the Zionist cause. Altogether, it has planted 250 million trees across a quarter-million acres in the Negev and elsewhere.

The Yatir Forest is not just about pressing an Israeli footprint into the desert sand. It is also claimed to have environmental benefits: holding back the desert, recharging soils with moisture, preventing floods in Be'er Sheva and, especially, fighting climate change by capturing carbon dioxide from the desert air. Alon Tal of the department of public policy at Tel Aviv University says that the Yatir Forest has, in an area with only ten inches of rain a year, "dramatically increased the carbon content of the soil and produced a lovely series of parklands."

To explore this further, I met up early one morning with researchers from the Weizmann Institute of Science, who have for twenty years been monitoring the forest. PhD student Yakir Preisler took me on one of his weekly trips. We drove from the institute's headquarters in Rehovot through irrigated fields and increasingly arid pastures towards the forest, which is set right against the fence separating Israel from the West Bank. This is military as well as ecological territory. The building from which the foresters work is known locally as the fortress. But Preisler is an independent researcher and we headed on to his mobile lab in a fenced enclave of the forest.

He showed me his measuring devices, strapped to hundreds of trees in the forest. They check the trees' growth and metabolism, including sap flow, leaf temperature and rates of photosynthesis and transpiration. A tower rising a little above the trees carries instruments to measure the forest's breath, like a low-level version of the giant tower above the Amazon outside Manaus.[19]

The forest is a monoculture of Aleppo pine trees, which are naturally found in wetter regions of the Mediterranean. They have had to adapt to the harsh desert conditions here. "They behave differently from those in Greece, for instance," Preisler said. They "grow like crazy" in the short spring wet season, when water is available for photosynthesis, then shut down for the rest of the year. With an eye to global warming, he predicts, "This is how forests will be in southern Europe in a few decades."

Such adaptability has its limits, however. If the pines are at the end of their range in Israel right now, will they survive in the coming decades? A year-long drought in 2010 killed ten percent of the Yatir trees, Preisler said, with more than eighty percent lost in places.[20] We looked at one of the barren areas left behind. Almost a decade on, there were few signs of new saplings growing. So it seems unlikely that, without new planting, the forest will survive beyond the lifetime of its founding trees.

At least for now the forest is trapping carbon and keeping the land cool. Or so I imagined. Back at the Weizmann Institute of Science, Dan Yakir, Preisler's boss and a specialist on the forest's relationship with the atmosphere, had a different take. He had calculated that the Yatir Forest has so far, if anything, caused warming. The growing trees are certainly taking up carbon dioxide. The KKL–JNF may be right that, when fully grown, each tree in the forest should eventually store some two-thirds of a ton of carbon.[21] They also transpire, albeit briefly, during the short spring growing season. That adds an occasional dose of further cooling. The problem is albedo. On the edge of the Negev, the dark foliage of the Yatir Forest canopy is replacing a light-colored, reflective desert surface. So the canopy is absorbing more solar radiation than the desert sand. The energy radiates into the air as heat, causing local atmospheric warming. It mimics what happens in Siberia in winter, where the contrast is between forest canopy and white snow.

So which effect is dominant? This calculation is complicated by the fact that albedo warming is local and constant, whereas the impact of the carbon take-up by the trees is global and cumulative. In the early years, warming is bound to be dominant. But barring disasters, every year that they survive and grow, the trees will store more carbon and deliver more cooling for the planet. There should be a crossover point, where the cumulative cooling exceeds the steady warming. So, has the forest passed it yet? Yakir says not. He estimates that it will take a total of eighty years of tree growth in the Yatir Forest before the cooling overtakes the warming. In other words, not until the 2040s, assuming that the trees survive that long. Which they may well not, since climate models predict that the Negev will experience declining rainfall and more frequent droughts, which could kill the forest and return its carbon to the air.

Yakir is no heretic. His work on the Yatir Forest has for many years been cited by Israel's forest lobby. It earned him the Israel Prize for Earth Sciences in 2019, the year of my visit. So this new calculation is something of a shock for his fans.

The story of the Yatir Forest will add to mounting concern about the wisdom of planting trees in places where they don't naturally grow. The concerns are ecological as well as climatic, and are held by a surprising number of Israeli environmentalists. They say that the trees their fellow countrymen love to plant in the name of greening their land are actually obliterating rare local ecosystems. "Most of Israel is a shrubland ecosystem, which is a high-value area for biodiversity. It must be preserved, not carpeted in forest," said Shofet at the SPNI. He grew up in New York before moving to Israel and says, "I have a deep desire to get American Jews to stop planting trees here."

He called in his colleague, director of biodiversity Alon Rothschild, to give me the lowdown. The Yatir Forest sits where desert and Mediterranean ecosystems meet, a region that contains "some of the most threatened habitats and species in Israel," he said.[22] He had a list. The planting of the Yatir Forest had reduced habitat for the dark-brown iris; a rare locally endemic daffodil called *Allium kollmannianum*; and the most southerly population of a species of wild wheat. Several ground-nesting birds, such as the pin-tailed sandgrouse, and shrub nesters like the spectacled warbler were in decline too, preyed upon by crows and jays that keep watch from the trees. Trees also take the sun needed by reptiles such as the endemic Be'er Sheva fringe-fingered lizards. Many of these species are officially protected. But Rothschild claimed that the KKL–JNF, which declined my request for comment, is rarely called to account for damaging their habitats.

This isn't just an issue for deserts. We should always be wary about planting trees where there have been few before, such as natural grasslands. Tree lovers may say, "It's only grass," but from the Brazilian Cerrado to the parks of South Africa and the arid shrubs on the edge of the Negev, grasslands are ecological gold dust. They are also rich carbon stores in their own right.

A quarter-billion acres of mostly grassy ecosystems in Africa have been targeted for forest restoration under the Bonn Challenge, based on what William Bond of the University of Cape Town complains is "the erroneous assumption that these biomes are deforested and degraded."[23] Plants and animal species that prefer open, well-lit environments will be pushed out, says Kate Parr of the University of Liverpool.[24] She means lions and elephants, bison and wildebeest. Africa's savannah grasslands are also working lands for 300 million people and store up to thirty percent of the world's soil carbon.

In other places, as we saw in China, trees planted in the wrong place can change local hydrology. Yes, they will transpire moisture into the air, so maintaining flying rivers and ensuring rain downwind. Yes, with luck they will bind together soils with their roots, regulate water runoff into rivers, hold back floodwaters and maintain river flows in droughts. There is an inevitable cost, however, with less water for rivers and underground water reserves that may be vital for ecosystems and humans. The lesson is that planting trees where there were none before requires extreme care and is often best avoided. They may do more environmental harm than good.

So, if not by more planting across the world's deserts and grasslands, how should we regreen our planet? In most places, most of the time, the best answer is not to plant at all.

# —16—

# LET THEM GROW

## *Only Nature Can Plant a Trillion Trees*

HERE IS A STORY you probably haven't heard about the Amazon. I certainly hadn't. Amid the deforestation, huge areas of the rainforest are regrowing. Naturally. In early 2020, Yunxia Wang, a geographer at the University of Leeds, reported that more and more of the clearance and burning there is not, as most news reporting suggests, of primary or intact forest. Most of the trees now disappearing from the forest are young, just a few years old. They are the result of regrowth in recently deforested areas. That finding does not diminish the terrible scale of the deforestation. But it does speak to nature's powers of recovery. And it illustrates how, even in the extreme conditions of modern Amazon deforestation, the forest is attempting to do what it has done so many times before after human invasion: regrow. It underlines the fact that, given the slightest chance, nature will reclaim its territory.

Using new Brazilian mapping data, Wang showed that in 2014, the last date for which she had detailed data, seventy-two percent of the forest being cleared was recent regrowth—double the proportion in 2000.[1] The typical time since the last clearing was just six years, and year by year the time was shortening. Probably some of the land was being deforested for a third, fourth or even fifth time since the carnage began in earnest half a century ago.

On the face of it, this seems very strange. Why would cattle ranchers clear forest, then let it regrow and clear it again? Well, it fits what we know about rainforest soils. Once cleared of forest and seeded with grass for cattle pastures, they quickly become exhausted. Grass won't grow after a couple of years. So ranchers abandon them and move on to clear more forest. What Wang measured for the first time was the extent to which, even with exhausted soils, the abandoned pastures recovered their trees. The regrowth won't instantly turn into mature rainforests—that takes time—so nobody should deny the threat posed by today's unprecedented ecological holocaust in the world's largest rainforest. What Wang's data show, however, is that the forest's powers for rapid recovery are much greater than often assumed. Policymakers may also appreciate Wang's observation that if Brazil wanted to revert to its pre-Bolsonaro promise to restore thirty million acres of forest land by 2030, it could probably achieve the target quite easily and cheaply just by making sure that secondary forests were allowed to regrow naturally in the Amazon. No planting required.

What is happening in the Amazon should be central to figuring out a path for global reforestation. For all the talk of trillion-tree programs, of Bonn Challenge pledges and of carbon offsets, it is becoming clear that there is another way to restore the world's forests: simply let them grow. As we saw in an earlier chapter, Philip Curtis of the University of Arkansas has estimated that only about a quarter of forest loss results in permanent human takeover of the land.[2] Much of the rest—whether trees are removed by forest fires, shifting cultivation or logging—at least has the potential for natural recovery.

The Global Partnership on Forest and Landscape Restoration, a network of academics and others set up to support the Bonn

Challenge, estimates that the world contains something like five billion acres of degraded land, an area the size of Australia. Of that, it calculates a quarter could in future regrow into closed forest and the remainder is suitable for "mosaic" restoration, in which forests are embedded into agricultural landscapes. Much of this forest is already of great ecological value and is regrowing right now. The UN estimates that at any one time there is as much as two billion acres of former rainforest in natural rehab. Richard Houghton of the Woodwell Climate Research Center in Massachusetts calculates that if those forests were left to regrow, they could capture up to three billion tons of carbon every year for as much as sixty years, potentially "providing a bridge to a fossil-fuel-free world."[3]

Where is all this natural regrowth going on? We heard earlier about abandoned farms in Europe and North America. The tropics contain a lot too, even as deforestation continues elsewhere. This regrowth can take many forms and may not pass muster as "intact" or as a replica of what was once there. Of course, it is often happening inside protected areas. But it is much wider than that. It is all-pervasive, constantly hiding in plain sight.

As I reflect on my travels around the world, seeking out places where forest activity is somehow special, I realize that I have often been driving right past the green shoots of the great forest recovery without even noticing. It is happening in the grass verges and abandoned pastures, in scrubby bush and forest margins. People dismiss it as "bush" or "undergrowth"; not proper forest. This is mistaken. It is the stuff of proper forests that are undergoing what ecologists call "succession." It may look like an afterthought of nature, but it is the driving force. If we humans suddenly departed this planet, it would be what nature did next. Nature abhors an ecological vacuum and hastily fills it.

Take the little-known success story of the Caribbean island of Puerto Rico. It is, says Thomas Rudel of Rutgers University, "proportionately the largest . . . forest recovery anywhere in the world."[4]

You have probably never heard of it. Even many forest scientists are in the dark. The story begins in a familiar way. Before the Spaniards showed up at the end of the fifteenth century, Puerto Rico's native seafaring people left their forests for the most part intact. But in the ensuing centuries, the colonists replaced the trees with sugar plantations in the lowlands and coffee in the central mountains. Sugar production peaked after the US took control following the Spanish–American War in 1898. By the mid-twentieth century, tree cover was just nine percent.

Then export markets for sugar and coffee declined and the plantations were abandoned. Meanwhile a US program to turn the island into a cheap-labor workshop, known as Operation Bootstrap, successfully persuaded smallholder farmers and herders to leave the countryside for urban factories, abandoning their fields. Subsequently, people also began migrating to the US. In this depopulated landscape, trees returned most quickly to the highland coffee fields, but later spread to the sugar plantations and pastures.[5] By the century's end, forest cover had returned to around half. Nobody had planted any of it. It seems Operation Bootstrap did more for nature than it did for the people of Puerto Rico.

There are many other examples of rural exodus and natural recovery, not least among Puerto Rico's neighbors. Small but densely populated El Salvador was once seen "as a Malthusian parable of population and ecological catastrophe," says Susanna Hecht of the University of California, Los Angeles.[6] Then, during the country's civil war in the 1980s, areas abandoned because of the conflict were taken back by forests. After their return in the 1990s, villagers in the Cutumayo basin decided to protect and continue the restoration of their forests.[7] As a result, a region that had eighteen-percent forest cover in 1978 had, by 2004, returned to sixty-one-percent cover. Elsewhere, Hecht says, the pressure to reclaim the forests was reduced by the exodus of many people to the US. They sent home money, so those left behind no longer had to grow their own food. Nationally, there has been an increase in forest cover of around forty percent.

While regrowth is still losing out to deforestation in much of the Amazon, Brazil's other great forest, the dry tropical Atlantic Forest of eastern Brazil, has also shown modest recovery. At its low point, just twelve percent remained. Most of it was cleared for coffee and cattle. Then plantation companies put in pine and eucalyptus trees to supply the pulp and paper needs of São Paulo and other cities. Suzano, the world's largest producer of pulp from eucalyptus, has hundreds of thousands of acres. More recently, there has been a switch to native species, often under the aegis of the Atlantic Forest Restoration Pact, a collaboration between government and landowners that aims to restore a third of the old forest.

Behind all the investment, hoped-for profits and promised jobs, natural regeneration has and will do most of the regreening in the Atlantic Forest, says Camila Rezende of the Federal University of Rio de Janeiro.[8] The past piecemeal destruction means that there are lots of patches of abandoned land close to forest remnants that can reseed and recolonize the abandoned land. Since 1996, some 6.7 million acres of the Atlantic Forest have regenerated naturally, she says. Most of it is within a few hundred feet of existing forests. Often this regrowth knits together forest fragments to create larger forested areas. This secondary growth now makes up almost a tenth of the total forest cover. That area could double by 2035, she thinks.

Some ecologists dismiss the virtue of degraded and secondary forests, arguing that secondary forests will never recapture what has been lost.[9] But, as we saw earlier, they do harbor much biodiversity and hold considerable amounts of carbon.[10] If allowed to regrow, they will hold much more of both. What may look like degraded or abandoned land is actually the cutting edge of a global reforestation. The first detailed attempt to map the potential for carbon capture from natural forest restoration found that regrowth rates assumed in reports from the UN's Intergovernmental Panel on Climate Change had been underestimated by a third. In the rush to set up planting projects to slow climate change, the power of nature to regenerate without our assistance had been seriously underestimated, Susan Cook-Patton of The Nature Conservancy, the study's chief author, told me.[11]

Ecologically, natural regrowth is almost always better than planting, says Robin Chazdon of the University of Connecticut, a staunch advocate of natural forest regeneration whose book *Second Growth* is a bible for its practice.[12] "Allowing nature to choose which species predominate... allows for local adaptation and higher functional diversity," she says. Moreover, forests recover "remarkably fast," with eighty percent of their species richness back in twenty years and frequently a hundred percent in fifty years.[13] Nature may choose something different from the past, especially if conditions have changed, but that is a sign of nature's dynamism and resilience, to be encouraged rather than thwarted. Karen Holl of the University of California, Santa Cruz, agrees: "In some highly degraded lands we will need to plant trees, but that should be the last option, since it is the most expensive and often is not successful."[14]

The evidence all around us is that nature has great ability to restore forests unaided. Most of the fifteen-percent increase in forest cover across the eastern United States in the past four decades has come from natural regeneration rather than planting.[15] That won't always work. But in places where there are no nearby trees to provide the necessary seeds for the restored forest, an obvious option is to plant a "founder stand," a small grove of trees that is left to reseed the surrounding area.

Devoted tree planters sometimes suggest that climate change will scupper these plans, and prevent regrowth of native tree species. But others say that natural regeneration will more often encourage growth of the local tree species best able to thrive in a new environment. "Natural regeneration helps species to shift and adapt to climate change," says Rebecca Wrigley of Rewilding Britain.[16]

Planting typically costs thousands of dollars per acre, once the expense of nurseries, soil preparation, seeding, thinning and the rest is included. Why bother if nature does it all better? That was the conclusion of Renato Crouzeilles from the Federal University of Rio de Janeiro, who analyzed reports on more than a hundred tropical forest restoration projects of all kinds around the world. Nature provided better canopy cover, tree density, biodiversity and forest structure.[17] Some supplementary "enrichment" planting might be handy to protect endangered species, he said. But however clever the foresters were, the planted trees were less well suited to the space they were occupying than those chosen by nature. And often, even with the best intentions, planting fails. The NGO Wetlands International, which for many years paid for communities to plant mangroves to protect coastlines, has concluded that most such projects fail and that improving conditions for natural regeneration, such as reducing coastal erosion, is the best approach.[18]

Nature's choices are unpredictable, of course. If circumstances change, so will nature. In Puerto Rico, many of the trees that took back control of abandoned plantations were not those native to the island. The first movers were African tulip trees, natives of the dry forests of tropical Africa. They had been widely grown in gardens and backyards across the island for their showy orange flowers. Then, given the opportunity, they took to the abandoned farms and spread rapidly, while native species failed to gain a foothold.

I first wrote about the strange story of Puerto Rico's forest recovery in a book on alien species. It asked whether we should condemn these ecological renegades, when their colonizing skills were often just what nature needed to jump-start recovery from human destructiveness. But things are changing on the island now. The aliens have improved the soils so much that native species have been returning.[19] Ultimately the natives may triumph, thanks to hurricanes. When Hurricane Georges ripped through the island in 1998, it obliterated a lot of tulip trees, which are not adapted to the strong winds of the Caribbean.[20] In the aftermath, native species gained ground. It is expected that the same phenomenon will have occurred on an even bigger scale after Hurricane Maria in 2017, which downed a quarter of the island's trees.[21] No census has yet established if the alien species there suffered more than the natives. We shall see in the long term what nature decides.

⸙

ALMOST WHEREVER WE MAKE ROOM for them, trees will return. Even intense pollution or radiation may be no barrier. Witness the twenty-mile exclusion zone around the Chernobyl nuclear site in northern Ukraine. After the disaster there in 1986, the government in Kiev decided that a thorough cleanup of the radioactivity in the thick forests and marshlands around the plant was impossible. So it settled on a policy of containment. It would leave the radioactive bogs and forests inside the zone to their fate. Nature accepted the challenge.

Very close to the stricken reactor all living things were obliterated. A pine forest right over the fence died within hours. Even the earthworms disappeared. Elsewhere, however, in areas of the exclusion zone where human habitation will be banned for centuries to come, the pines have recovered and prospered. They are invading abandoned villages and now cover two-thirds of the zone. They harbor a growing population of wild animals.

Sergey Gaschak, scientific director of the Ukraine government's Chernobyl Radioecology Center, drove me down an overgrown lane to the wrecked remains of Buryakovka, one of the 113 villages evacuated after the accident. Fallout maps showed it to be a radiation hotspot. But he had set up two cameras whose shutters were tripped by passing animals. He took out the camera memory cards and put them into his laptop. Up popped a procession of Przewalski's horses—part of a herd released into the zone in the 1990s—plus foxes, moose, wild boar, pine martens, wolves, racoons, badgers and a couple of rutting male red deer. The exclusion zone has more bears and wolves now than in neighboring protected areas that are not radioactive, he said. They don't actively like radioactivity, but they do like the absence of people. The radioactive forests were hosting what Gaschak called "a reinstatement of nature unique in Europe."

Such uninhabitable wastelands are still, thankfully, exceptional. But what they show, once again, is nature's resilience and adaptability. As Chazdon puts it, "Forests constantly renew themselves." Places like Chernobyl also suggest that by constantly referring back to a supposedly "intact" past, we are in danger of short-changing that natural dynamism. To dismiss forests that are in the early stages of developing into rich mature ecosystems as degraded, purely on the basis of their past condition, is surely perverse. Yet, Chazdon says, that approach has become commonplace among many conservationists. Regrowing forests "continue to be misunderstood, understudied and unappreciated for what they are—young self-organizing forest ecosystems that are undergoing construction," she says.[22]

The world's forests will be restored not by trying to recreate the past, but by providing the space for such forests to find their own new future. Sometimes that will happen within still-empty former forest lands and within forests often dismissed as "degraded" wasteland. Sometimes—and just as valuably for both nature and the climate—it will happen in agricultural landscapes, which we will explore next.

— 17 —

# AGROFORESTS

## *Farmers as Part of the Solution*

FORTY YEARS AGO, the Sahel region of Africa was a byword for drought, hunger and environmental apocalypse. Today, the region is still dry, but some of its driest plains are covered in trees. Satellites passing over the fragile land on the southern flank of the Sahara Desert now often see green rather than yellow. In the fields below, farmers say crops grow better among the trees, soils are more fertile and their people continue to be fed despite one of the fastest-growing population rates in the world. Most impressive of all, once again nobody planted the trees. After years of ripping them from the ground, communities just stopped ripping and let them grow.

This story starts in one village, Dan Saga, more than ten miles from the nearest highway in the Maradi region of southern Niger. The village suffered dreadfully from drought in the 1970s and 1980s. Crops repeatedly failed. But in 1985, two young men returning from jobs in mines over the border in Nigeria began a transformation that is still playing out in Niger and adjacent countries. The men were coming home to plant crops before the short rainy season began. They were late, however, and by the time they arrived home the rains were already underway. Their neighbors had followed government advice to clear their fields of weeds, tree roots and small bushes so that the soil would be easier to plow. In their rush to get next year's crops in the ground, the two latecomers abandoned the clearing and planted in among the roots, hoping for the best.

To their surprise, their harvest was better than their neighbors'. Intrigued, they did the same the next year, with the same result. Clearly, trees were good for crops, and their fellow villagers took note. From then on, the farmers of Dan Saga left roots in the ground and allowed stems to grow from stumps. The trees might have got in the way of plowing, but they improved the soils, capturing moisture and fixing nitrogen. Before long, the trees were also yielding firewood and leafy animal fodder, while providing shade and blocking desert winds. There was no looking back. Subsequent analysis of satellite images showed that from virtually no trees in 1984, Dan Saga had an average of twenty-four trees per acre by 2005; then forty-four by 2012 and fifty-nine by 2016.

News spread across Maradi, a region almost the size of Switzerland, and villagers began to ignore the government advice, which had been in place since colonial times. The people of Zinder to the east and Tahoua to the west did the same. Today, some two hundred million trees have sprouted on more than a million family farms across more than seventeen million acres of southern Niger and neighboring Mali.

The history of West and North Africa over the past sixty years is littered with failed efforts to halt the spread of the desert by planting trees where there were none before. Algeria in the 1970s constructed a thousand-mile "green dam" of Aleppo pine, modeled no doubt on Israel's efforts with the same species.[1] It failed. Mauritania launched a "green belt" around its capital, Nouakchott, between 1975 and 1990, and then did it all over again in 2010. The latest attempt is Africa's equivalent of China's Great Green Wall, known to its Francophone sponsors as la Grande Muraille Verte. The aim is to plant a cordon of trees about ten miles wide for more than five thousand miles from Senegal and Mauritania on the Atlantic coast, through Niger, to Djibouti on the Red Sea. "Over the next ten years, more than 120 million acres of land will be restored, which will help sequester an estimated 250 million tons of carbon," declared the African Union's launching press release in Paris in 2015.[2] Some $8 billion has already been pledged for the project. By 2021, UN agencies claimed it would restore a quarter-billion acres, double the original target. The project, which had been stifled by lack of donor funding, had been hitched to the newly announced UN Decade on Ecosystem Restoration, and the job would be done by 2030.[3] It was "a historic opportunity to conserve biodiversity, address climate change and enhance food security simultaneously."

Well, maybe. Many opinion-formers backed the scheme. Including the journal *Nature*, which ran an editorial calling for the world to "get Africa's green wall back on track."[4] But few desert specialists believe it will work. Chris Reij of the World Resources Institute notes, perhaps with a hint of exaggeration, that "if all the trees that had been planted in the Sahara since the 1980s had survived, it would look like Amazonia. But eighty percent of them have died." But, by contrast, the revival of trees in southern Niger demonstrably does work. The trees do not die. Because all that is happening is that farmers are nurturing the roots of trees already in the ground. The trees are naturally adapted to the region; they are growing naturally; and they are being tended by the local farmers. It is a recipe for success, not expensive grandiloquent failure.

The man who brought this dryland revolution to the attention of the outside world was Reij, then a geographer at the Free University Amsterdam. A decade ago he told me, "In 2004, I drove five hundred miles east from the capital, Niamey, and I thought: bloody hell, there are trees everywhere. I had never seen such densities on cultivated land. You couldn't see villages because they were hidden behind trees." It was, he said, a total transformation since his first visit to Niger two decades before. "It was obvious they took a lot of pride in their achievements. Every tree had been pruned so it would develop a trunk."[5]

Reij decided to set up a study, using remote-sensing images collected by Gray Tappan of the US Geological Survey. They have both kept a close watch on developments since. Overall, the trees are estimated to have captured from the atmosphere more than thirty million tons of carbon.[6] Not that the villagers care much about that. What they see is trees improving their livelihoods. They get more sorghum and millet to feed their families, and gain an extra income of typically $1,000 per year from selling wood, fodder, fruit, pods and leaves. The cash is vital in dissuading young farmers from migrating to find work. Tappan says that the denser the human population in an area, the more trees there are.

The trees are also community assets. Dan Saga now has a village committee that organizes collecting and planting seeds to increase tree density. It has imposed rules that prevent certain trees from being felled and set up night patrols to prevent illegal cutting.[7] It also encourages sustainable use of the trees. There is a regular firewood market. One of their leaders, Ali Neino, runs training programs for other villages.

The farmers have learned to favor particular species. They especially prize a tree variously known as white acacia, winterthorn or, in the local Hausa language, the gao tree. It is squat and thorny, with a trunk up to six feet in diameter, and roots as deep as the tree is high. It can survive drought and if necessary hibernate underground for decades. Its flowers attract bees, its roots fix nitrogen in the soil and its bark makes medicines. Average millet yields can double beneath the shade of the gao tree. Not bad, even before you count the leaves as fodder for livestock and the wood for making everything from canoes to fires. No wonder old creation myths in the Sahel call the gao the "tree of life."

There is, it should be said, a different version of the story of the rediscovery of the benefits of trees here. It involves an Australian missionary and agronomist called Tony Rinaudo, who told me he came up with the idea when he was working in Maradi in the 1980s. While encouraging villagers to plant trees to "hold back the desert," he discovered that many village soils contained an "underground forest" of root systems that would sprout if left undisturbed. He says that during the drought of 1984, a year before the Dan Saga story, he was offering food to farmers if they would nurture these trees. The handful who did ended up with better crops.

His current employers, World Vision Australia, promote Rinaudo as "the forest maker."[8] But he told me, "It was a farmer-led phenomenon. I played a facilitator role only." He had never worked in Dan Saga, he said. Maybe they heard about his initiative, or maybe they came up with the idea independently.

Whatever the precise history, lots of development experts have now adopted the practice. It has a fancy title: Farmer Managed Natural Regeneration (FMNR). The Niger government teaches it, and has its own "FMNR ambassador," Tougiani Abasse from the National Institute of Agricultural Research. UN drylands ambassador Dennis Garrity says that among agriculturalists they are "passing a tipping point, where the idea of natural regeneration is really beginning to snowball."[9] Well, it only took thirty-five years for the message to get from Dan Saga to the UN experts!

"In some respects FMNR isn't new at all," says Rinaudo. "It is simply an adaptation of ancient practices that have been going on around the world for centuries. I have seen it in East Timor and Burundi. In Malawi, there are over 2.5 million acres of FMNR, and no evidence of outside influence." Europeans have a variant method of getting trees to sprout from their stumps. They call it coppicing. But what makes Niger stand out are the dire circumstances under which people adopted the idea of allowing trees to grow; how contrary it was to the wisdom of government experts at the time; and the sheer expanse and speed of its spread. "The important thing," says Rinaudo, "is that it is accessible and usable by the most resource-poor farmers, independent of any power or external aid."

FOR A LONG TIME—despite its obvious visible presence on the land—this grassroots revolution in the Sahel remained largely invisible to outsiders. Only a few researchers spotted that something was happening. I remember talking in the early 1990s to British geographer Michael Mortimore, who said farmers in southern Niger were feeding fast-growing populations by harvesting rainwater and raising trees. But his passion was another revolution going on two thousand miles to the east, in a district of Kenya a couple of hours' drive from Nairobi. In a book he co-wrote called *More People, Less Erosion*, he referred to it as the "Machakos miracle."[10] It too had a narrative of locals ignoring colonial expectations.

Back in the 1950s, British administrators described the hilly cattle-herding district of Machakos as an environmental disaster: overpopulated, treeless, eroded and turning to desert. By 1990, it had seen a sixfold increase in population. Yet the doom-mongers had been proved wrong. Rather than turning to desert, Machakos had many more trees than before, less erosion and a vibrant local economy.

The local Akamba people had saved themselves by switching from herding to farming, and by digging terraces on hillsides, capturing rains and planting local trees everywhere they could. When I visited, they were not just feeding their families, but selling vegetables and milk to Nairobi, green beans to Britain, avocados to France, mangoes and oranges to the Middle East and cut flowers to the Dutch. "Farmers here grow maize to eat, but everything else for cash," Benjamin Ikombo of the nearby National Dryland Farming Research Station told me as we inspected mangoes and pawpaws on his own small farm on a Machakos hillside.

When I first reported on the Machakos miracle, development professionals told me it was a myth, or at best a one-off. The more I looked, however, the more examples I found of communities in arid regions using trees to heal their land. In the highlands of western Kenya, the Luo people were even more densely packed than the Akamba. But amid their fields they were planting woodlands that produced timber, honey and medicines.

A young Swedish forest researcher, Peter Holmgren—who later became director of the Center for International Forestry Research—found that, "contrary to prevailing beliefs," there had been "a rapid increase of planted woody biomass" in Kenya. Planting had peaked between 1986 and 1992, he said, a time when the country's birth rate was the highest in the world, with an average of seven children born to every woman. "Population density is positively correlated with the volume of planted woody biomass," he stated.[11] That was another real jab at conventional environmental thinking. The seemingly inevitable link between overpopulation and environmental degradation just wasn't happening. As Tappan had noticed in Niger, more densely populated places tended to have more trees, not fewer.

It seems to be a global trend. Far from banishing trees from their land in an effort to boost yields, many farmers are nurturing them. This, in the lingo, is agroforestry, and it has its own international research center. Robert Zomer of the World Agroforestry Centre in Nairobi says that getting on for half of the world's agricultural land has at least ten-percent tree cover. That represents 2.5 billion acres of land—an area the size of China.[12] The figure has been increasing by three percent per decade. The Machakos story seems to be not so much a local miracle as a global phenomenon. Where there are more people, they tend the land better and there are often more trees as a result. In the booming economies of Southeast Asia, even as natural forests continue to be lost, tree density is increasing fast in agricultural areas and around settlements.[13]

AGROFORESTRY CAN TAKE many forms. The trees may help sustain farmers' crops, or be key crops in their own right. Farmers grow trees such as coffee, cacao (which makes chocolate) and vanilla as part of complex garden systems that rely on the shade of the forest canopy. When grown on plantations, these crops cause deforestation. Notoriously, cacao has caused widespread forest losses in the two biggest-producing countries, Ivory Coast and Ghana. But in Cameroon, which is another of the top six, smallholders plant cacao beneath the forest canopy.

Some years back, I met one of them. Joseph Essissima was quite old. He showed me the trees he had planted in the near darkness beneath the forest's trees sixty years before. His farm was dimly lit, but full of life. Amid the cacao, he had planted other tree crops, and nurtured native trees for their fruit, bark, timber and medicinal leaves. I recognized oranges, mangoes, avocados and cherries. I asked about another tree. "We keep this because it attracts caterpillars that we eat. They are very tasty," he said. Essissima was not unusual. His neighbors did the same. A study of these agroforests found that farmers typically planted twenty-five other tree species among their cacao, and their agroforests occupied sixty percent of the farmland in the area.[14]

Essissima told me that environmentalists had once condemned him and his fellow agroforesters for invading the forest, but now they came to praise him for protecting it. It wasn't his methods that had changed but their perspective. With me on the visit was a researcher from the International Institute of Tropical Agriculture, Jim Gockowski. "These cacao plantations have more than half the species you would find in a natural forest," he said. "It may not be virgin forest, but if the farmers didn't plant cacao here, they would be planting maize or palm oil, or turning the land over to cattle pasture." They were, he said, "managing one of the most biologically diverse land-use systems in Africa."

Poor farmers in tropical countries are often criticized for clearing forests. Yet many see trees as a vital part of their farms, providing fruit, timber, medicinal plants or fodder for their livestock; fertilizing soils, shading fields and harboring insects that eat pests or pollinate the crops.[15] Their trees may be no substitute for unfenced intact forests. Still, often they form genuine ecosystems able to evolve and grow in complexity, mimicking the ecological "succession" of natural forests, says Robin Chazdon of the University of Connecticut.

Sometimes they add to biodiversity, even in existing forested areas. When James Robson of the University of Manitoba set off to investigate what happened as Oaxacan farmers abandoned their fields and orchards to work in Mexican cities, he expected to see a revival of biodiversity as the forest returned. But to his surprise, species numbers declined on the abandoned land. He says the complex mosaic of forests and small-scale fields that the farmers had created in the forest harbored a zone of hyper-biodiversity that disappeared when the farmers left, and the old forest returned.[16]

That may be exceptional, but it is clear that intact ecosystems and protected areas need hinterlands of farms with scraps of nature within them. If they become islands in a sea of intense land use, they lose biodiversity, says Claire Kremen of the University of California, Berkeley: "Even the largest protected areas will lose species over the long term, unless surrounding landscapes can be managed to provide connectivity."[17] The threat can only increase in an age of changing climate. Many species will need living corridors down which to migrate if they are to keep up with shifting climate zones. Farms, especially agroforests, can provide those corridors.

There is perhaps a wider debate here between what we may call the sparers and sharers. The sparers believe that the best way of saving forests and other intact ecosystems is to intensify farming on land already cleared. The sharers believe nature is best served by integrating forest with farming, through agroforestry. It is a tricky debate to resolve. You can see the sparers' case. Norman Borlaug, the famed father of the Green Revolution of high-yielding crop varieties, declared half a century ago that his crops would save nature as well as feed the world and reduce rural poverty.[18] He argued that farmers could grow more crops on existing fields, rather than invade the forests. But, as we have seen, biodiversity cannot thrive in islands of intact nature. Even the bigger protected forests need to be connected to wider landscapes if they are to thrive.

There are other factors of equal importance in real landscapes in the real world. For one thing, intensive does not necessarily mean more productive. Certainly, the size of a farm is no guide. Smallholders, with plots of less than five acres, produce about a third of the world's food on under a quarter of the agricultural area, according to Vincent Ricciardi of the University of British Columbia.[19] Such smallholdings seem to have both a bigger output and a smaller environmental footprint.

Economically, it is far from clear that even very productive high-tech farming will spare more land for nature. The world's food system is ultimately a market. So intensification often just increases profitability, which raises the incentives for farmers to clear more forest. "Borlaug thought if you addressed poverty in the forest border, they'd stop taking their machetes into the forest. Actually, they get enough money to buy a chainsaw and do much more damage," says Tony Simons, director of the World Agroforestry Centre.[20] Look at Brazil. The cleared Amazon rainforest is not feeding Brazilians; it is exporting beef and soy for the profit of a few. All over the world, says Thomas Rudel of Rutgers University, countries don't farm less land as their crop yields increase—they farm more.[21]

In the real world, the choice between sparing and sharing will rarely be as stark as in a seminar room. But what is clear is that agroforestry is an increasingly popular idea among everyone from smallholder farmers to agricultural scientists, ecologists, governments and development banks. The governments of sub-Saharan Africa in particular see it as a way of combining environmental and development goals. It makes sense, given that eighty-three percent of Africa's population live and depend directly on the land, and by some estimates two-thirds of that land is in some manner "degraded" for want of trees.

Backed by a billion dollars from the World Bank, the African Forest Landscape Restoration Initiative aims to recruit agroforestry systems to restore tree cover to a quarter-billion acres by 2030. Among the enthusiasts, Burkina Faso has promised 12.5 million acres, Malawi and Tanzania a similar amount and the Democratic Republic of the Congo twenty million acres. Through Farmer Managed Natural Regeneration, Niger's farmers have already achieved similar coverage.

Many governments also hope that agroforestry can help them meet both their Bonn Challenge reforesting targets and their Paris climate pledges. Agroforests typically hold six times more carbon than monoculture plantations, though only a seventh as much as natural forests, according to Simon Lewis of the University of Leeds.[22] Aid agencies are keen too. Retaining trees on farmland is the core of another major program, organized by the World Agroforestry Centre and funded by the EU, called Regreening Africa.[23] At the heart of this project—constantly repeated in its literature—is the concept of Farmer Managed Natural Regeneration. Niger veteran Rinaudo is involved.

Chris Reij has seen many such high-flown projects come and go over the years. "Pledging is easy; implementing is a lot harder," he says. "A lot of this will never happen, even when they claim it has." So where is he putting his time and enthusiasm? His answer is Ethiopia. Since the disastrous drought in the 1980s, farming communities there have given their labor to planting 2.5 million acres of trees in Tigray province, which was the epicenter of the famine. Most of the trees are on common land, fenced to keep animals from browsing and farmers from plowing.[24] The trees were intended primarily to supply wood for fuel and construction, taking pressure off natural forests. But they also regreen the landscape, moderating temperatures, recycling rainfall, maintaining soils and harboring biodiversity.

Unlike the farmers' actions in Niger, this return of the trees is directed from the top—as a patriotic endeavor. Prime Minister

Abiy Ahmed claimed that his farmers planted a world-record 350 million trees on a single day in July 2019. He promised five billion more in 2020, with a total of twenty billion by 2024—enough to meet half of Ethiopia's Paris pledge. Critics say this program is socially coercive. Some twenty million villagers have to spend a month or more every year on planting, yet lack clear individual or community rights to benefit from their labors. The planting can be ecologically questionable too, both because it will deluge the landscape with foreign eucalyptus and acacia trees and because the survival rates of the planted trees are often low.[25]

Even so, where it works, it can work well. It is only a generation ago that people in the Tigrayan village of Abrha Weatsbha were on the verge of abandoning their land as it turned to sand. Community leader Aba Hawi has been singled out as the architect of an ecological recovery that has involved planting trees, constructing small dams to catch rainwater, propagating fruit trees and tending beehives. The scene today is certainly a sharp contrast to the famine of the 1980s.[26] Back then, the BBC was reporting a "biblical" holocaust in an area very close to the village. Today there are raised water tables, reduced soil erosion and increased farm productivity—and trees have been at the heart of that transformation.

IF FARMING AMONG TREES is good, then surely it is time to take a fresh look at one of the most contentious topics in forest conservation: shifting cultivation. For thousands of years, farmers have roamed the forests, making temporary clearings to grow crops. Rather than putting trees onto their farms, they put farms into their forests. For a long time, environmentalists have dismissed such farmers as forest destroyers. Their techniques have been denigrated as "slash-and-burn" farming. I admit that I have used the term myself. The farmers have been seen as evil, ignorant or—at best—left with no choice other than to trash their environment.

They were "tradition-bound peasants, trapped by ignorance and unable to manage their resources properly," as Thomas Tomich, formerly of the World Agroforestry Centre and now at the University of California, Davis, once put it sarcastically.[27] To this day, WWF sometimes condemns the practice. In June 2020, its website still claimed that "slash-and-burn fires... don't just remove trees; they kill and displace wildlife, alter water cycles and soil fertility, and endanger the lives and livelihoods of local communities."[28]

Many former critics have changed tack, however. They regard the practice as often highly productive and ecologically sustainable. As Robin Chazdon of the University of Connecticut argues, traditional shifting cultivators are not slash-and-burn monsters but masters of managing forest ecology. Their languages are often much richer in detailed description of the stages of a forest's growth than the lexicon of Western forest ecologists, she notes. Charles M. Peters of the New York Botanical Garden says of shifting cultivation that when "properly done it is a really amazing solution to enriching sterile tropical soils."[29]

Many such cultivators are reluctant to show off their activities, believing that they are still despised by outsiders. Not so Patrick Gomes, whom I met in the forested hills of southern Guyana. He offered to take me to see his patch, so we drove his four-by-four out of the village and up into the hills, waving to two women riding a bullock to forage in the woods and greeting a man out hunting with a bow and arrows. We parked the vehicle and entered the forest, stepping warily across a stream where Gomes had recently found an anaconda digesting its prey. Then we walked down a track, where he pointed out the many plants he had put into the ground over the years. They included wild bananas, medicinal plants and some straight-stemmed plants that he said were still used to make hunting arrows.

Finally, we reached a clearing of about a couple of acres. The first field was planted with cassava, a root crop that provides flour for bread, can be fermented to brew beer and makes a popular spicy sauce used like ketchup. The second field had a cash crop of peanuts. What I saw was no anarchic assault on the forest. Gomes had a system. He explained how he came from his village regularly to tend his crops, plant trees or clear fresh land. This plot would grow crops for a year or perhaps two, until the fertility of the soils began to decline. Then, after the last harvest, he would plant some valuable trees for future use and move on, leaving the land fallow to recover its fertility.

The forest regrew swiftly here. He pointed out a field he had abandoned the previous year. It already contained rich undergrowth, some of it above head height. "I leave an area fallow for fifteen to twenty years," he said. "By then it is indistinguishable from the canopy forest around."

Nature didn't seem to have a problem with his invasion. As we walked back, two macaws rose into the sky. There were no anacondas to waylay us, but the forest was alive with sound.

## PART IV

# FOREST COMMONS

Foresters and environmentalists alike often forget that their beloved forests are inhabited—usually by people who know the forests much better than outsiders like them. Whether shifting cultivators or traditional pharmacists, Indigenous tribes or other folks, tree huggers or even artisan loggers with chainsaws, these people are the forests' best custodians and most sophisticated managers. We explore their world, hear their stories and slay some myths about the "tragedy of the commons." For the world is learning that forest commoners and the trees they tend provide the route—and the roots—to the great restoration.

# —18—

# INDIGENOUS DEFENDERS

## *Why Tribes Do Conservation Better Than Conservationists*

ON THE FACE OF IT, Tessa Felix was like any young office worker. She turned up promptly for our morning meeting, cycling from her home in trousers and a T-shirt bearing a corporate logo; she logged on to the internet connection as we chatted and cursed its slow response. But when I asked about her work, her world suddenly seemed very different from that of the average office drudge. She began telling me about her grandfather, who was once a chief, or *toshao*, of the village where we met—a village deep in the south of Guyana in South America. "When I was a child, my grandfather took me into the forest. He went fishing. He showed me the secret places of our people and told me our stories," she said. "My work now is about preserving my grandfather's legacy. I feel the land belongs to me and my people. Everything we want is here. My grandfather said that I was born from the Earth and I believe that. Our land is like our mother. Now I want to fight for our land."

We were in Rupununi, a region of Guyanese forests and savannah grasslands bordered by Brazil to the west and south and Suriname to the east. Felix lives in Shulinab, a village of some five hundred people, two hours' drive down a potholed road from the rodeo town of Lethem. Her people are the Wapichan, a tribe of Amerindians. They are forest hunters with bows and arrows, but also farmers and cattle ranchers. They are, like many Indigenous communities, trying to embrace two worlds. They speak their native language, but also fluent English. They can spend weeks in their territories, walking the ancient trails, swimming the creeks, climbing the trees and visiting their ancestral graves and sacred forests. But they navigate as much with GPS apps on their smartphones as by traditional knowledge and would happily drive home in one of their beaten-up all-terrain vehicles.

Guyana is one of the most forested countries in the world. It is eighty-seven-percent trees. Its government gets international plaudits for its low rate of deforestation. But most of the forests are in areas populated and used by Indigenous communities. Of these, the traditional territory of the Wapichan people is the largest. To an outsider, their land can appear empty. Traveling between villages, I drove for hours across grasslands and past forest without seeing more than an occasional hunter or cattle herder. There are just nine thousand Wapichan, occupying a traditional territory the size of Wales.[1] That amounts to just over a square mile per person. Some question whether, in the modern world, they can rightfully claim so much land. The Wapichan insist that they have a right to full legal title.

"Our land is being taken away from us often without us even knowing," Nicholas Fredericks, once a precocious Wapichan cowboy and now a village leader and coordinator of land-use monitoring, had insisted the night before as we cracked open beers while Brazilian soccer played on the village's solar-powered TV. "These forests are our life, but they are being taken from us. We have to protect them for the future of our people."

He shares that task with Felix, who, as a worker for the community, investigates reports of their territory being invaded—by gold miners crossing the border from Brazil, and by loggers, illegal fishers and cattle rustlers. Her last trip, she told me, took her to the Marudi Mountains in the south of the Wapichan lands, to record how a new Canadian-owned mine was polluting rivers that the Wapichan rely on for fish. "I know villagers who get their fish from that river," she told me in horror, showing on her phone the video she took of a once-clear rushing stream turned to a cloudy cascade of mud and mercury. On one six-day patrol along the river that forms the border with Brazil, Wapichan scouts found thirty illicit paths crossing the border, six of them in regular use by miners. The patrols have a deterrent effect. "The rustlers fear what they call the 'monitors with smartphones.' They turn back if they hear we are around."

The Wapichan are proud of their adoption of technology, and see no reason why it should make their land needs and entitlements any the less. Felix—a twenty-five-year-old IT whiz, political sophisticate and bush tracker—is a living refutation of the idea that the new generation has to choose between modern and traditional ways. She wants her traditional lands *and* a better internet connection.

Nobody knows how long the Wapichan have been in Rupununi. They probably moved north from the Amazon before the eighteenth century. Like most Amerindian communities, they suffered both from raids by slave traders and from European diseases. In the nineteenth century, Jesuit missionaries established schools and churches, and persuaded them out of the forests. In the early twentieth century, British colonial authorities created a series of "reservations" around major villages and offered up the rest of the Wapichan traditional lands for commercial cattle ranching. There were not many takers, though a Scottish entrepreneur leased 1.6 million acres of what he claimed was the most remote ranch in the world.[2] It once had a cattle trail all the way to Georgetown, on the coast. The trail was abandoned long ago, but the ranch is still there, albeit with only a sixth of the cattle.

As the British left in the 1960s, Guyana's Amerindians campaigned to get their traditional lands formally recognized in the independence settlement. Felix's grandfather was among the elders who made the claim to Queen Elizabeth. They were granted title to less than a sixth of their claim, however. When, in the 1990s, they demanded the rest, the country's president, Cheddi Jagan, challenged them to say how they would use the land. So they decided to map and document their existing use and draw up future plans.

The task took almost a decade, with financial and technical support from, among others, the EU, the UN Development Programme and the British government. Tom Griffiths of the UK-based Forest Peoples Programme, which works with the Wapichan, says that probably no Indigenous community in the world has devoted so much time and energy to such a project.

"In the early days there was no GPS," said Tony James, former chief of all the Wapichan. "We just walked, following government maps and adding detail to them from our knowledge. But with GPS we have been able to do things much more precisely." Angelbert Johnny, a former village chief, remembered, "The elders and experts on the creeks, forests and mountains were our guides as we walked, or took boats and bicycles and horses to survey the land. We sometimes traveled for a month. It was hard. People got bitten by snakes. One guy got lost for two days. But we went everywhere and mapped everything. Often we had to put our GPS equipment on poles and push them up through the forest canopy to get a signal."

They ended up with forty thousand digital points collated and cross-checked by Californian digital mappers. They found numerous errors in the government's maps. Several of their seventeen villages were wrongly situated and different creeks endlessly confused. Government mappers have refused to accept their errors, says Ron James, who was in charge of the mapping. "They just say our maps are wrong, even when the evidence is obvious and corroborated by satellites."

Along with the cartography, the Wapichan documented their lives, culture and traditions. Claudine La Rose, from Shulinab village, recalled, "The elders told us how we came to be in the mountains, about the sacred sites and the spirit grandfathers that preside over natural resources. How, if you cut down certain trees in the forest you will get sick and die, punished by the spirits who live in the mountains and creeks." They also developed a plan to protect their land using traditional knowledge and methods. In particular, they want to create a community forest, managed for hunting and gathering, for shifting cultivation and, they hope, for science and tourism. Covering 3.5 million acres, it would be one of the world's largest community forests. They reasoned that if they documented their existing customary use, the government should recognize their land. Things weren't so straightforward, however. It took seven years for the government even to agree to begin talks, and four years later, as I write, there has been no outcome.[3]

During my visit, I joined a two-day meeting of Wapichan leaders and activists at which they discussed what to do next. It was held in the nursery school in Marora Naawa village, where shifting cultivator Patrick Gomes, whom we met in the previous chapter, was chief. Portraits of the country's president and prime minister hung on the classroom walls. People had traveled for a day or more to be there. There was anger that government officials had told three village chiefs not to attend because the meeting was not sanctioned and was therefore illegal. Villages would suffer if their leaders attended, they said. Such threats were real. Activists said government services that might be taken for granted elsewhere only reached Wapichan villages in the form of high-profile gifts and favors from the president and ministries. Things such as drugs for clinics, solar panels to provide electricity and the internet service now available in three villages came through this patronage. Most villages feared losing these amenities.

The Wapichan claim not only that they have a right to their land, but also that their knowledge of it means they are the only effective custodians. They hunt for deer, bush hogs, agouti and armadillos, but only take what they need for their families. They know that killing more would empty the larder for future generations. They practice shifting cultivation in the forests. Their cattle range over huge unfenced areas. But they leave some areas of land entirely alone, either because of its spiritual meaning or as refuges for wildlife.

Today there are probably fewer commercial activities in the forests than there have been for a century. The Wapichan used to bleed the native bully trees for *balata*, a latex that was a major export for Guyana until the 1970s. There were airstrips to fly it out. Today the trade has ended and the grass strips are maintained only for occasional use by the flying doctor. Ranching is no longer a thriving business because transport costs are too high. Many families send members to work in Georgetown, in Brazilian gold mines or in Boa Vista, the boom town just over the border.

If they are to prosper, the Wapichan badly need new sources of income from their lands. For a while, back in the 1990s, it seemed as if they might earn substantial royalties by allowing their traditional medicines to be turned into new pharmaceuticals for the outside world. One of their number, Conrad Gorinsky, the clever son of a Polish cattle rancher and an Indigenous mother, had left for a biomedical career in Britain, taking with him plants containing two natural potions that he had known as a child in the forests.[4]

The Wapichan had long used the bark of the greenheart tree to prevent malaria. In his lab, Gorinsky isolated the active ingredient. He did the same with a poison found in the leaves of a local shrub that he and other Wapichan children had made into spitballs and thrown into ponds, where it made fish so disoriented that they could easily be picked out of the water by hand. He figured that the powerful nerve stimulant in the leaves could have many medicinal uses. But the hopes of turning these potions into cash evaporated. Gorinsky and the Wapichan fell out over who should have the patent rights and no drug companies took up his discoveries.

Today, the best chance to raise Wapichan incomes may lie in nature tourism. Currently, the few hundred visitors to their territories each year are mostly scientists, students and ecotourists. Many stay at the big commercial ranch. Some Wapichan villages have built small guesthouses, modeled on successful enterprises run by other Amerindian communities. But, while the food is good, they rarely meet Western expectations. One I stayed in required a walk of three hundred feet in the dark to find a latrine. There were rattlesnakes, I was told.

Such drawbacks can be fixed, however, and the daytime experience is magical. There is spectacular bird life, with harpy eagles, pearl kites, savannah hawks, the near-endemic Finsch's euphonia and a small bright-orange finch called the red siskin. Once thought to be extinct, the red siskin was rediscovered by Wapichan rangers, prompting WWF to organize a major biodiversity assessment of Rupununi. Their rangers led the investigators to more than a thousand species. The Wapichan believe international recognition of the bird life could encourage visits by birdwatchers and bolster their land claim. An obvious tourist destination would be the forested Kanuku Mountains in the north of the Wapichan territories. The mountains have more biodiversity than anywhere else in Guyana, with black caiman, giant river otters, giant anteaters and arapaima, the largest freshwater fish in South America.[5] The Guyanese government eyes the mountains as a protected area to place under its control, but a more logical approach would be to open it up to tourists under the environmental stewardship of the people who know it best, the Wapichan.

We should not be too romantic. A clash of cultures is being played out in Wapichan communities. Self-sufficiency and dependence on the forests are losing out to computer games, canned vegetables and the lure of cash. Many Wapichan young people seek work and education outside. But villagers still slaughter their own animals for meat and make cassava and tapioca bread from local ingredients, as well as wine, jams and a pungent pepper sauce—usually sold at village stores in old vodka bottles. They make their own leather too and every village has a sewing circle, where women make traditional hammocks and baby slings.

Tony James told me of his plans to introduce summer classes in bushcraft for children who go away to secondary schools. "We should teach them how to make bows and arrows, how to survive in the forest and how to hunt," he said. "At the end of the course, they could go into the forest on their own and prove their prowess in finding food. And then return with what they have hunted or gathered to prepare a village feast. The young women could welcome the hunters. There could be a bushcraft graduation ceremony. We could even fly in tourists to watch."

Is that wishful thinking? Perhaps, though I found the sense of purpose among the community's leaders deeply impressive. Yes, the modern world poses threats, but it also offers opportunities for them and their forests. Just maybe, modern technology and the global links it can bring will take them on a road where digital mapping, environmental concern and global advocacy can secure their land rights for future generations—and secure their forests.

EARLIER IN THIS BOOK, I stood at the boundary between the forested Xingu Indigenous reserve and the Tanguro soybean farm in Mato Grosso. It was a sobering experience to see two worlds collide so dramatically on the ground. It was almost equally staggering to view on a computer screen back home satellite images of that boundary extending for hundreds of miles, and to see the same pattern repeated across the entire Amazon basin. The Amazon today is a giant patchwork made up of huge islands of green forest surrounded by brown areas of land cleared for farming. In places, satellite time series show the brown advancing, year by year. In others, the green holds firm. Why the difference? Because the secure green areas mostly belong to assertive Indigenous communities. In most places, at most times, they are the ones who are saving the Amazon.

The Xingu reserve, which is home to a number of Indigenous groups, covers 6.4 million acres. It is eighty-five-percent forested, but surrounded by a sea of soy. Immediately to the north the Kayapo people control more than 27 million acres of forest. There is no deforestation within their territory either. Together the two reserves form a green island rather larger than England, surrounded by the brown of cattle pasture and soybean farms.[6]

Those communities, numbering just a few thousand inhabitants, have held back an invasion by outsiders that has engulfed areas close by. Where government agencies have failed to act, the communities have had to violently repel loggers, gold miners, cattle ranchers and soybean farmers. On one famous occasion, after learning that gold miners had moved into a remote area of their reserve, Kayapo warriors raided the camp by boat and on foot, destroying the equipment and successfully pressuring the government to send in helicopters to remove the miners. The miners never returned.

We used to be told that forests have to be saved from the people who live in them, but the opposite is usually the case. The three hundred or so Indigenous territories created in the Brazilian Amazon since 1980 are now widely recognized to have played a key role in a dramatic decline in deforestation this century. The laws creating the reserves give Indigenous communities the right to repel outsiders and control of the natural resources their reserves contain. They have used those powers. One study found a two-thirds reduction in deforestation since 1982 in parts of the Brazilian Amazon covered by the "collective property rights" of Indigenous communities.[7]

Another found that across the Brazilian Amazon from 2000 to 2014, deforestation was at 0.6 percent per year in areas under the control of Indigenous groups, compared to seven percent outside.[8] This is not, incidentally, because their land has fewer people than other parts of the forest. On the contrary, population density is often higher inside Indigenous territories than outside, says Rodrigo Begotti of the University of East Anglia, "dispelling the often-repeated argument that there is too much land for too few Indians."[9] To the west, in the Peruvian Amazon, the story is similar. Where Indigenous communities have the title to their land, it "reduces clearing by more than three-quarters and forest disturbance by roughly two-thirds," says Allen Blackman of the Washington think tank Resources for the Future.[10]

Forest dwellers are modern-day forest saviors and they do best when the going is toughest, says Christoph Nolte, now at Boston University. He found that in the Brazilian Amazon, "indigenous lands were particularly effective at avoiding deforestation in locations with high deforestation pressures."[11] You can see that too in the satellite images. State-protected areas may be overwhelmed; places where the locals are in charge rarely are.

Let's not see Indigenous groups as pure environmentalists. They are defending their land and resources as much as they are defending nature. For them the two are indivisible. In the Brazilian Amazon, an estimated seventy thousand people still make a living by tapping wild rubber and harvesting Brazil nuts, tonka beans and acai, a palm whose fruit is now considered to be the third most valuable forest product out of the Amazon, exceeded only by beef and timber. Much of this rainforest harvesting now happens within twenty or so extractive reserves set aside for local people.[12] They cover an area of Amazon jungle the size of England. The first reserve commemorates the work of Chico Mendes, the rubber tapper from the western Amazon whose campaign to protect forests by creating such reserves was adopted by Western environmental groups in the 1980s. He was assassinated by a local rancher. But his disciple, Marina Silva, later became environment minister, and secured his legacy by establishing a network of reserves that protect the forests while using them.

"The Amazon is a working landscape, protected by those who use it; not a model of purity," says environmental historian Susanna Hecht of the University of California, Los Angeles. Even if that sometimes disappoints lovers of intact forests, they should remember that the "intact" Amazon is mostly regrowth of pre-Columbian forest gardens. In any case, they should see that forest communities, their techniques and jurisdiction, are part of the solution, not the problem.

The Amazon is not exceptional. In critical frontier zones across the world, two things are becoming clear. The first is that forest communities protect the forests; the second is that they are generally better at it than governments or outside environmentalists, especially when they have secure title to their land. (The World Resources Institute reports that in Ghana, which has one of the world's fastest recent rises in loss of primary forests, two-thirds of that loss has been inside formally protected areas.)[13] This is not just about forest cover, extent or even carbon. Forests managed by Indigenous communities also turn out to have more biodiverse ecosystems than areas protected by national parks and other reserves, according to an analysis of more than fifteen thousand locations, from Brazil to Australia, led by Richard Schuster, of the University of British Columbia. There was little doubt, he concluded, that "it's the land-management practices of many Indigenous communities that are keeping species numbers high."[14]

Often, in tropical forests, it is the areas with a history of human occupation that are the richest in species.[15] As a former head of the UN Environment Programme, Klaus Töpfer, once put it, there is "clear and growing evidence of a link between cultural diversity and biodiversity, between reverence for the land and a location and a breadth of often unique and special plants and animals."[16] Or, as Guatemalan Indigenous activist and 1992 Nobel Peace Prize winner Rigoberta Menchú has it, "It is not accidental that where Indigenous peoples live is where the greatest biological diversity exists too."

Such analysis is still a shock to many environmentalists, who have for decades encouraged governments to remove Indigenous peoples and local communities from forests in the name of environmental protection. Over the decades more than forty thousand people have been displaced from nine protected areas in six Central African countries, with many more deprived of hunting and gathering grounds.[17] The global total may run into millions of people removed, says anthropologist Dan Brockington, now at the University of Sheffield, in a strategy he calls "fortress conservation."[18]

Around seventeen percent of the land area of the planet is now "protected" for nature. About half of these protected areas overlap the traditional territories of Indigenous peoples, while many more—probably most—impinge on areas claimed by non-Indigenous rural communities. Yet according to the UN Environment Programme, most of the world's national parks still do not have any form of community management. This is both unjust and stupid.

In one of many cases cataloged by human rights activists, the Wildlife Conservation Society (WCS) in the US convinced the government of the Republic of the Congo in 1993 to turn a million acres of forest in the northwest of the country into the Nouabalé-Ndoki National Park. In the process several thousand semi-nomadic Bayaka "pygmy" people were expelled from their ancestral hunting grounds in the park. It is an old story, but cannot be consigned to history. The injustice remains. Cruelly, the park is adjacent to a large logging concession, also annexed from Bayaka land. In effect, the loggers and conservationists have carved up the forest, cutting out the Indigenous inhabitants.[19] To rub salt into the wounds, in early 2020 the WCS's website was still describing the park as "arguably the best example of an intact forest ecosystem remaining in the Congo Basin," having "little or no contact with people."[20] No mention of the Bayaka.

The WCS is not alone in its troubling history, or its failure to apologize or make reparations. In 2020, the UN Development Programme (UNDP) cataloged how WWF has for years been funding guards in another Congolese green fortress, the Messok-Dja National Park. These guards have beaten up and intimidated hundreds of Bayaka, expelling them from their ancestral lands, imprisoning them, burning their houses and confiscating their crops. The UNDP reported that the top echelons of the environmental organization had, despite public denials, known for years about the abuse there and in other parks that WWF has funded in neighboring Cameroon, the Central African Republic and the Democratic Republic of the Congo.[21]

Environmentalists often talk about engaging with forest communities. But "funds from well-meaning people in Europe and North America are still helping to pay for these violations of indigenous peoples' rights," says John Nelson of the Forest Peoples Programme.[22] Stephen Corry, the director of Survival International, which helped record the Congo abuse for the UNDP, calls it "a disaster for conservation, as it's destroying the very people who are the best conservationists."[23] It was also beginning to impact on funding. In the aftermath of the UNDP report, US authorities halted funding of more than $12 million for WWF, WCS and others working in the region.[24]

There is a culture war going on here over what being green is about. Douglas Sheil of the Norwegian University of Life Sciences says conservationists often suffer from a "tainted-nature delusion: the view that nature is only good enough to conserve when it satisfies our mental ideals"—that is, a nature without people.[25] But this is also, more crudely, a turf war over who is in charge of the wild lands. Is it the locals who live in and among the forests? Or is it the outside experts, who arrive in SUVs to assess the ecology and depart again to write up their reports? Perhaps that conflict helps explain why it has taken so long for many environmentalists to recognize that Indigenous peoples may know better than them how to manage forests.

Even so, a shift in perspective is coming. I was fascinated by an interview in *Yale Environment 360* with top forest botanist Charles

M. Peters of the New York Botanical Garden.[26] He was publicizing his 2018 book *Managing the Wild*. I expected an exposition of how Western science is deployed to understand and conserve the forests. Instead, he argued that "local people know a lot more about how to manage tropical forests than we do" and have "this incredible body of traditional knowledge." Of a visit to Dayak people in Borneo, he enthused, "These guys are managing 150 species of tree in a hectare [2.5 acres]. We Western foresters can't manage four species in a plot... As a forester you just go: oh my God, how do they do this?"

It is no longer acceptable to see such folk knowledge as some quirky leftover from ancient times, charming but essentially irrelevant to the modern world. Indigenous knowledge is in most respects superior to the knowledge of outsiders. If we are to save, nurture and restore the world's forests, the best expertise for achieving that is alive and well, living in those forests right now. They cannot be bystanders in the conservation of their territories. They should—and must—be in charge of the process.

THIS NEW NARRATIVE HAS growing force in public discussions about conservation in the tropics, where central government is often weak and corrupt. But the hard lesson that people rooted in the land usually know better how to manage it than outsiders holds true for elsewhere. Listen to Valérie Courtois, a forester and director of the Indigenous Leadership Initiative, which fosters native stewardship across Canada.[27] "I can't think of one landscape in Canada that hasn't been affected in some way by humans. It's a false premise to think that landscapes are at their best without us," she said in a 2018 interview. An Innu, whose whole lifestyle revolves around following migratory woodland caribou herds, she rejects the idea of keeping people barricaded off from nature. "As Indigenous peoples, we are a part of that biodiversity... This idea of absence is totally artificial." The Innu want their land back, and many environmentalists support them.

In the old-growth forests of British Columbia, home to around a quarter-million Indigenous peoples in almost two hundred First Nations communities, Indigenous people and environmentalists have been joining hands to protect the temperate rainforests since the 1990s. That was when George Watts, a young leader from the Tseshaht Nation, formed a tribal council to represent the fourteen Indigenous nations on Vancouver Island to conduct what he called the "war in the woods." They blocked roads, halted logging and forced the government to address long-standing Indigenous land demands.

More than six hundred primarily Indigenous communities are in Canada's boreal region, but they control only two percent of the country's forested lands—a fraction that is smaller than in most developing countries.[28] But where they are in charge of forests, the results can be spectacular. The Xaxli'p people resumed control under the terms of a community forest project in 2011. It covers sixty thousand acres of Fountain Valley, north of Vancouver, and has developed its own forest recovery plan, aimed at recouping what has been lost since industrial loggers invaded in the 1950s.

The plan pledges to "manage the forest for ecological and cultural sustainability, based on the needs of the Xaxli'p community."[29] That means restoring lands that have either been clear-cut or left exceedingly dense through fire suppression. The restored forests contain open, patchy and biodiverse woodland fit for moose grazing, while protecting water sources and with resilience to wildfires. It is work in progress, but the direction is clear. "By having Xaxli'p elders and a forest ecologist work together, the community was able to develop mutual understanding and a common language, which effectively linked traditional and Western scientific knowledge," says Sibyl Diver of Stanford University, who has studied the history of the project.[30]

Disputes over the fate of British Columbia's forests are far from over. When environmental protesters blockaded logging roads on

Vancouver Island in summer 2021, demanding that a deferral of logging be turned into permanent protection, they brought criticism from some Indigenous community members, who said the uninvited "third party activists" were not welcome on their territory. Still, it is clear that mutual understanding is possible. And that applies south of the forty-ninth parallel too. Numerous studies are showing that Indigenous peoples across the US are superior stewards of the land. Donald Waller of the University of Wisconsin and Nicholas Reo of Dartmouth College spent years investigating the ecological state of forests under the charge of the Ojibwe and Menominee people in their reservations between Lakes Superior and Michigan in northern Wisconsin, and comparing them with both historical records and state and private forests on surrounding territories. The pair found that the tribal forests are "more mature, with higher tree volumes, higher rates of tree regeneration, more plant diversity, and fewer invasive species."[31] Not because the forests were unused, but because they were used better.

The two tribal groups successfully combine professional standards of forestry with their own knowledge and cultural norms. They have done this "for more than a century, demonstrating cultural resilience despite strong historical pressures to alter their management," said Waller and Roe.

The Ojibwe have based forest management of their three reservations—Bad River, Lac du Flambeau and Lac Courte Oreilles—on restored cultural traditions, the most important and distinctive of which is respect for the wolves. These apex predators share their land, and hunt deer, just like their human neighbors. But rather than seeing them as competitors, the Ojibwe see them as collaborators in controlling deer numbers. Keeping the deer under control improves the survival rate of many tree species, as well as benefiting wolves, the wider forest ecosystem and of course the Ojibwe themselves—ensuring, for instance, supplies of birch bark to make the canoes for which they are famous.

The Menominee too have a reputation for wise management. Their 234,000-acre Menominee Forest is famously visible from space as a dark-green block of maple and aspen, birch and hemlock, pine and oak. It is surrounded by dairy pastures carved from the forests by German immigrant farmers who have occupied much of the area since the nineteenth century. But throughout that time, the Menominee people, many of whom live in the forest, have pioneered their own Indigenous methods of sustainable forest management, and fought off successive efforts by outsiders to tell them how to manage their forest.

Their philosophy is "managing the forest for a long time," says Marshall Pecore, who has himself been in his post as forest manager for thirty years.[32] Some hemlock and white pine trees in his forest are more than two hundred years old, and rise two hundred feet into the sky. Yet conservation and production go hand in hand. The Menominee like to say that, since they established their management system in 1860, they have harvested the forest's volume twice over, yet have more tree cover than when they started out.

Their methods are rigorously scientific. Pecore and his colleagues maintain a continuous inventory of their trees, each of which is numbered and regularly measured. Only a settled proportion of the older trees, with a diameter of ten inches or more, are harvested. But this scientific approach is built on their cultural attachment to the forest. In a recent interview, Menominee community activist Guy Reiter told journalist Alexandra Tempus: "If you live in a forest, you pay attention to things. That's better than any paper or research. You've got to watch it; you've got to be part of it."[33] The result has been hailed by the NGO American Forests as "one of the most historically significant working forests in the world."[34]

Waller agrees. As he puts it, the Menominee and Ojibwe "can teach us something important about what we need to learn about managing forest land." Public forest managers should stop telling the tribal foresters what to do and start learning from them.

— 19 —

# COMMUNITY FORESTS

## *A Triumph of the Commons*

A GROUP OF WOMEN gathered together in the sun on the edge of the Piple-Pokhara community forest, amid the green foothills of the Himalayas in central Nepal in early 2018. They settled down to an afternoon of making wooden handicrafts, a new enterprise they had recently started. They had good reason to admire their handiwork. Not only had they created the toys and other wooden items, but the forest from which the wood came was a product of their industry too.[1]

"Twenty-five years ago, the trees here were so sparse you could see people walking on the distant hills," said Ramhari Chaulagai, the chair of the committee in charge of the Piple-Pokhara forest. The area had been blighted by deforestation. With no trees, the hills were dry. "Our people had to go a long way to get to water." So, in the 1990s, the local communities formed a forest users' group and began planting trees and nurturing natural regrowth on the bare hills. It was a collective effort to create a community forest over which, under then-new government legislation, they could assert their control. Across the world in the Amazon, you could still claim land by clearing forest; in Nepal, you claimed it by planting and nurturing trees.

Now, a generation on, the inhabitants of communities around the forest, in Makawanpur district southwest of Kathmandu, can gather wood for fuel and handicrafts, graze their animals in the shade of the trees and take walks under the canopy of the forest. Homnath Gautam said he had planted many of the alder, pine and fir trees in the forest. Now retired, he still climbed the hill behind his house every day, admiring the natural regeneration taking hold and checking out the wildlife: leopards, porcupines, pangolins and deer.[2]

In the past quarter century, Nepal has developed a successful system for community management of its forests in the foothills of the Himalayas, some of the most densely populated parts of the country.[3]

It has been an amazing turnaround. Until the 1990s, trees were owned by the government and corrupt officials made Nepal a byword for rapid deforestation. Now, some 22,000 autonomous community forest users' groups have legal rights to manage and control access to around a third of the nation's forests, covering more than five million acres.[4] Millions of people are able to use, profit from and simply enjoy them. National forest cover has increased by a fifth to forty-five percent, one of the fastest rates of recovery in the world, with most of the growth in community forests.[5] The people's forests suffer fewer fires and less illegal felling. Soils have improved. Watersheds suffer fewer landslides and springs are reviving. New community leaders have also emerged, as grassroots democracy has flourished, according to analysis by Helvetas, the Swiss development agency, and poverty has declined.[6]

The community forests operate as businesses, supplying local needs, whether tree shoots to feed livestock, wood for firewood, or fruit and medicines. With secure ownership, the communities have come to regard the forests as long-term assets, to be nurtured rather than stripped bare. As Johan Oldekop of the University of

Manchester concluded after a 2019 study, "Nepal proves that with secure rights to land, local communities can conserve resources and prevent environmental degradation."[7]

⁂

THE STORY OF NEPAL'S community forests, and many others like it, challenge a widespread view about community management of the planet's commons—whether forests or water supplies, pastures or fisheries—summed up in the phrase "the tragedy of the commons." The term was coined by American ecologist Garrett Hardin in a hugely influential paper published in 1968, in which he claimed that collective ownership of natural resources was a recipe for disaster.[8] He argued that with nobody in charge, everyone will grab as much as they can while they can. Individuals will all be acting "rationally": they will reason that because the resource will be trashed by others, they had better join in while they can. The result will be collective disaster. Trees will be logged without limit; underground water will be pumped dry; pastures will be grazed till they turn to desert.

The lesson usually drawn has been that nature should be taken out of collective control—which Hardin saw as no control at all—and put into the hands of either an assertive state or private companies with a stake in its long-term survival, and the ability to defend it.[9]

In fact, most of the real-world evidence disproves Hardin's hypothesis. His nemesis proved to be Elinor Ostrom of Indiana University. She read Hardin's paper and hit back, arguing that communities all over the world successfully manage their natural resources, through local agreement, even though neither the state nor private owners were involved.[10] She had seen this at work first in the community forests of Nepal. Her subsequent detailed analysis found the same among Spanish farmers sharing scarce water resources, herders on Japanese mountain meadows and Indonesian fishers. Even in Swiss cattle pastures, cheese producers had for centuries managed open pastures because this allowed cattle to graze more efficiently than in fenced private plots. Communities usually showed common sense, Ostrom said, and the worst thing to do was to "impose rules from the outside."

Ostrom's ideas were not universally appreciated. One reviewer, an academic economist, called her classic takedown of Hardin's ideas, *Governing the Commons*, "an evil book... because it contains a nasty vicious attack on private property rights, the linchpin of a civilized order."[11] But in 2009, "this perky, untidy figure"—as the *Economist* was to call her in an otherwise flattering obituary three years later—became the first woman to win the Nobel Prize in Economic Sciences. Since her death, evidence in support of her case has kept stacking up. Many more researchers with deep knowledge of local resource management tell similar stories.[12]

David Bray of Florida International University has spent a lifetime studying Mexico, where more than 2,300 rural communities have long-standing ownership of roughly sixty percent of the country's forests. It works, he says. They log carefully and selectively. Mexican communities are preserving millions of acres of forest cover "with little impact on flora or fauna biodiversity." Deforestation rates in community-owned forests are lower than in government-protected areas.[13]

For instance, the Zapotec Indigenous communities of the Sierra Norte mountains of Oaxaca have logged their pine and oak forests for many decades. They have their own sawmills and furniture workshops, but they also run tree nurseries. Of the half-million acres of land under their control, almost eighty percent is forested. Of that, two-thirds is logged and a third is zoned by the communities for conservation. The management is underpinned by strong community ties to neighborhoods, churches and schools, and obligations to provide up to twenty days of labor a year to collective tasks. Again, it works.[14]

Bray calls such systems "community capitalism." It is "done by communities where virtually no one has more than a primary-school education, [but where] community members talk to each other, build management institutions, establish rules, patrol resource boundaries, sanction people who break the rules, and under the right circumstances can manage the commons sustainably," he says. "When you give a community an incentive to cooperate—such as a valuable resource to manage—it is a powerful spur to collective action." He says archly that it is "something the world might want to emulate."

Mexico and Nepal are unusual in having community control of forests as the national norm. Most governments try to control forests themselves. According to the Washington-based Rights and Resources Initiative (RRI), forest communities, including Indigenous peoples, have legally recognized rights to just fifteen percent of the world's forests.[15] Latin America did best, but in sub-Saharan Africa the figure was just seven percent.[16] In practice, however, whatever the law says, probably around half of the world's forests are effectively occupied by traditional peoples, who use them to varying degrees for hunting, gathering and shifting cultivation. They make extensive and clever use of the resources of their forests. A study by the Center for International Forestry Research (CIFOR) found that harvested natural resources are the largest component of the incomes of people living in and around tropical forests—more than crop farming, wages or livestock. This forest fecundity was "largely invisible in national statistics," however, "because the produce is either consumed in the home or sold in local markets unmonitored by national data-collectors," said Frances Seymour, then CIFOR's director.

Forests also form an important part of culture, which is an important part of forest protection. For instance, almost all societies seem to have sacred groves of trees. I have never gone looking for them, but I have stumbled on them from time to time—in

Cambodia and South Africa, Ghana and India, Guyana and even Estonia, where a popular pre-Christian convention survives in which each village has its own sacred forest and people each year leave harvest gifts in the groves for their ancestors. Elsewhere in Europe, the Celts and Druids, Gauls and Lithuanians, Finns and Welsh all have histories of tending sacred groves.

In Africa, the Kikuyu in Kenya protect groves of the mugumo tree for prayers and animal sacrifices; in Sierra Leone, the groves harbor medicinal plants; in Ivory Coast they are used for initiation ceremonies. In Asia, the Mansi people of western Siberia maintain sacred groves in the boreal forests; the Japanese place Shinto shrines in woodlands; the Chinese have sacred tree-covered mountains. One study found more than fourteen thousand sacred groves in India alone. Australian Aboriginal people have "dreaming" groves; the Maya cultivated cacao, their "food of the gods," in them.

In modern times we still plant groves to honor the dead. Britain's main memorial ground for its armed forces and many others is a national arboretum with 25,000 trees. In California, there is the National AIDS Memorial Grove in San Francisco. The state's sequoia groves—the world's tallest, largest and oldest trees, grown and tended by local people over centuries—are now dedicated to past presidents. Closely related are forests used for "green" burials. In Germany, 45,000 deceased have "tree burials" each year. My son is buried beneath a tree in a green-burial plot at Brookwood cemetery in Surrey.

⁂

COMMUNITY MANAGEMENT OF the world's forests works staggeringly well when they have genuine control. Luciana Porter-Bolland of the Institute of Ecology in Mexico City found that state-protected areas are deforested on average four times faster than nearby community forests.[17] A review of 130 local studies in fourteen countries, conducted by the World Resources Institute (WRI) and the RRI, found that community-owned forests suffer less deforestation and fewer fires, while storing more carbon than other forests.[18] "No one has a stronger interest in the health of forests than the communities that depend on them for their livelihoods and culture," said RRI's Andy White. As Andrew Steer, the director of the WRI, told me once, "If you want to stop deforestation, give legal rights to communities." And if your concern is to halt climate change, then, according to Ashwini Chhatre of the University of Illinois, "We can increase carbon sequestration simply by transferring ownership of forests from governments to communities." After examining eighty forest "commons" across the tropics, Chhatre concluded, "When their tenure rights are safe, they conserve the biomass and carbon in such forests." But he also found that "when local users perceive insecurity in their rights (because the central government owns the forest land), they extract high levels of livelihood benefits from them."[19] The logic, he told me, "is very simple, but the policy implications are profound. We do sincerely hope that our paper will help overturn ideas about the tragedy of the commons."

This is happening. In 2006, India passed a Forest Rights Act that gave a quarter of the country's 1.4 billion citizens, those living in forested areas, the right to a section of community forest for their use, plus access to government funds to manage it. Implementation has been slow, but in 2013, the country's Supreme Court called the Act "an imperishable endowment . . . the largest ever land reform in India, and among the largest in the world."[20]

⁂

THERE ARE FEW clearer examples of how community forests can be superior to government forest "protection" than the state of the Maya Forest in northern Guatemala today. The government of Guatemala created the Maya Biosphere Reserve in 1990, to protect Central America's largest rainforest. At the time American conservationists advising the government felt betrayed that, along with a series of protected parks in the reserve, a big chunk was given to local communities for logging. They saw it as a lost chance to save the heart of what they had called the third most important biodiversity hotspot on the planet, home to more than fourteen hundred plant and 450 animal species, including jaguars, pumas, tapirs, spider monkeys, alligators, harpy eagles and macaws.[21]

How wrong could they be? Since then, illegal cattle ranches have been wrecking the national parks containing the forests in the west of the reserve. They have some of the fastest rates of deforestation in the world. Almost a third of the forest in the largest park, the 835,000-acre Laguna del Tigre National Park, has been lost since 2000. But in the east of the reserve, the once-denigrated community forests are still intact, a shining beacon of conservation covering more than 860,000 acres. Deforestation rates are a fraction of one percent. Together, they make up one of the world's largest and most successful community forest experiments.

Jennifer Devine of Texas State University, who spent years in the region before conducting a detailed review of satellite images, calculates that up to eighty-seven percent of the deforestation in the parks is the result of illegal cattle ranching. Her interviews with officials and activists on the ground found a near-unanimous view that at least two-thirds of the deforestation, including most of the large clearances for giant ranches, has been directly funded by criminal gangs trafficking drugs through the reserve. Her study shows the community forests to have been far more resilient to land grabs than the state-protected parks.[22]

The Maya Forest of the northern Guatemalan department of Petén is one of the great ecological jewels of Latin America. The forest was once controlled by the ancient Maya civilization and contains many of its archaeological remains, including the famous jungle pyramids. Areas cleared by the Maya swiftly regrew and for much of the time since the area has remained thinly populated, with perhaps five thousand inhabitants, most of them descendants of the Maya. Change came after a US-backed coup in 1954 brought to power a government that began shipping in landless peasants. A civil war followed that lasted into the late 1980s, when a new government sought a new future for the region by establishing the reserve, which is in total more than twice the size of Yellowstone.

The original plan, drawn up by the US government's aid agency, USAID, with help from US-based affiliates such as Conservation International and The Nature Conservancy, was to impose strict protection in the reserve.[23] But, with an estimated ninety thousand people by then within its boundaries, that proved impossible. There were mass protests. Meanwhile, grassroots organizations representing local forest users demanded the right to establish community forests, where they could continue harvesting timber and other forest products.

In the end, strict protection was restricted to national parks mostly in the west of the reserve. Eleven forest concessions, covering about a fifth of the reserve, were granted in the east. The compromise deal "was anathema to many of the conservationists involved," according to a retrospective analysis by John Nittler for USAID.[24] Some conservationists refused outright to deal with local communities, regarding them as "squatters," even though "most of Guatemala's protected areas were carved out of [their] territory," says Liza Grandia of the University of California, Davis.[25] But the squatters turned out to be the best forest protectors.

A decade ago, the thirty-mile road through the reserve to Carmelita village was flanked by forests all the way. Now it passes through a lawless and treeless landscape until it reaches the village's forest concession. Carmelita is a century-old community, originally established as a settlement for extractors of forest products. Its inhabitants today are ethnically mixed. Some are descendants of the Maya, while others arrived with the settlement programs of the mid-twentieth century. Its four hundred inhabitants control the first, and one of the largest, of the community forests, covering 133,000 acres. They specialize in adding value to the timber through carpentry and selling locally sourced and made furniture in their own local store.[26]

What lies behind their success in conserving their forests, when all around there is deforestation and lawlessness? One of the rules set by the government when establishing the concessions was that communities should use the forests sustainably. While laws of all kinds are routinely flouted in the state-protected areas, the people of Carmelita and the other concessions largely stick to the rules. They have all applied for, and received, certification of the sustainability of their harvesting methods from the WWF-backed Forest Stewardship Council (FSC). They are almost the only forests in Guatemala with that status.

The communities also benefit from long-term advice from the Rainforest Alliance, an American NGO, that helps them find markets for their products. These include valuable timbers such as mahogany and Spanish cedar, which despite its name is a New World tree, and several non-timber products from trees, some of which have been cultivated here since the time of the Maya civilization. Among them is chicle, a latex tapped from sapodilla trees. Chicle was exploited on a large scale in the early twentieth century here to make Wrigley's chewing gum, until the company found a synthetic substitute. Now it has a niche in organic and Fairtrade markets selling "natural" chewing gum, and is also FSC-certified. Other products include allspice, a tree berry found widely in the Maya Forest; xate palm leaves that are exported by the million to the US and Canada for use in floral bouquets on Palm Sunday; and Maya nuts.[27]

The Maya nut tree, a relative of the mulberry, is the largest tree in these forests, growing up to 150 feet tall. It is present in large numbers because it was extensively cultivated in the forest by the ancient Maya civilization. Its nuts, which were an important part of Maya diets, have recently undergone a revival. Across Central

America, women-run enterprises harvest, process and market the nuts, which are eaten raw, ground to make flour and made into salable products such as nut cookies and ice cream.

The people of these community forests are increasingly commercially savvy. Each of the concessions has its own business plan for sustainably harvesting their forests. Uaxactun—the largest, covering 200,000 acres—sells mahogany direct to US guitar makers. It also profits from tourists visiting nearby Maya ruins. The Rainforest Alliance estimates that, taken together, the concessions generate more than $6 million in annual revenue.[28]

At the heart of the community concessions is a strong collective organization, the Association of Forest Communities of Petén (ACOFOP), which has its origins in an old union of chicle tappers. "The forest is an economic asset to the people," ACOFOP's deputy director Juan Giron says. "If the person benefits from natural resources, he or she sees them as an asset. Land rights guarantee access to the forest and in the case of the forest concessions in Petén, this access leads us to take better care of these resources."

Devine's satellite images show that care. In an average year, just one percent of the many fires in the reserve occur within the community forests.[29] This, she says, is because the communities construct fire breaks and conduct constant patrols—augmented by drones and GPS trackers—that look for fires and protect their borders from outsiders. "Community foresters defend the forest they manage against encroachment," says Devine. They do it because their ownership of their forests "is the hard-won product of a two-decade struggle."

The struggle continues. Two of the eleven concessions collapsed after intimidation from ranchers who invaded their land, but the others thrive. One of the biggest, Cruce la Colorado, survived the assassination of its leader, David Salguero, shot dead outside his office after confronting ranchers deforesting within the concession. Current trends suggest that by the middle of the century almost all the surviving trees in the Maya Biosphere Reserve will be in community-run areas.[30]

The people who often get the blame for deforesting are the estimated fifteen thousand "illegal" settlers in the parks—many of whom have nowhere else to go. They live in very poor conditions and undoubtedly clear a few areas to grow crops or raise cattle, says Devine. There are regular high-profile evictions. But her analysis of some 4,500 satellite images of the deforested areas shows that they are not the real environmental villains. "What we see in the photos is very large cattle ranches that are often empty and far from settlements or crop cultivation." Their size and location mean they cannot be connected to the settlers. The images "confirm that cattle ranching and not subsistence farming is responsible for the majority of forest loss."

So, if the ranchers are the deforesters, who are the ranchers? Devine's interviews with locals who have detailed knowledge of the area reveal a near-unanimous view that most of the ranching is directly funded by what are known locally as *narco-ganadero*, or narco-ranchers. They are the Guatemalan allies of two rival Mexican cartels that are also implicated in the large-scale conversion of rainforests to cattle pasture over the border in Mexico's Calakmul Biosphere Reserve. They are untouchables, operating what some call a "bribe or bullet" tyranny. Faced with this choice, most park guards choose the bribes.

The traffickers buy the ranches and pay for the deforestation partly as a convenient means of laundering profits, says Kendra McSweeney of Ohio State University, and partly because the biosphere reserve occupies half of Guatemala's long border with Mexico, a border that has become the main route taken by cocaine traffickers to reach their markets in North America.[31]

Devine's images reveal that the ranches contain miles of clandestine roads leading to the border, and around a hundred small airstrips that are out of sight of the authorities and out of the hands of rival gangs. They also reveal that most of the ranches are surprisingly empty. This too, she says, is a sign of narco-ranchers. They barely bother to stock their giant ranches, which can cover hundreds of acres. They destroy the forests but barely use the land. There could hardly be a starker contrast with the other world inside the biosphere reserve.

Recognizing the success of the community concessions in protecting the reserve's forests, the Guatemalan government at the end of 2019 granted an extension to the original twenty-five-year license given to the Carmelita community. More renewals should follow, but threats remain. One comes from the US. As I write, a proposal is being discussed by the US Senate to offer the Guatemalan government $60 million to beef up security in the Mirador basin, a part of the reserve known for its Maya archaeological ruins, which American tourists love to visit. The bipartisan Mirador–Calakmul Basin Maya Security and Conservation Partnership Act, which claims the support of conservation groups and the Guatemalan Congress and president, was introduced by right-wing Republican senator Jim Inhofe.

On the face of it, a "security and conservation partnership" sounds like good news. But remember the 1990 plan to provide full protection for the entire biosphere reserve. That sounded good too. This plan risks being similarly counterproductive. One critic said that it would likely absorb more than half of the Uaxactun concession and up to a fifth of the Carmelita concession. The communities, who have not been consulted about the proposals, would lose control of their forests. Further timber harvesting, however sustainable, would be banned. So if this crazy project went ahead, the only people with a record of being able to repel the narco-ranchers would be sidelined. In truth, the project seems like a throwback to an era when naïve ideas about creating fortresses for conservation held sway and locals were the enemy. We know better now, though some governments are taking a while to catch up.

— 20 —

# AFRICAN LANDSCAPES

## *Taking Back Control*

THE MAJORITY OF THE WORLD'S most interesting forests take a while to get to. It had been a long journey from Dar es Salaam, the biggest city in Tanzania, on the western shore of the Indian Ocean. After a day's drive and then a three-hour boat ride through East Africa's biggest mangrove swamp, we finally arrived at Mfisini, a small village hidden among the mangroves that cover the delta of the Rufiji River. Mfisini was a model of order. Raised on a small area of dry land, the village had dozens of tall coconut trees dotted among the huts, each tree owned by a different household. There were solar panels on some roofs. We passed a couple of wells outside huts. In the center of the village we met Yusuph Salelie, the village committee chair, who escorted us into a wooden hut that served as his office and the village meeting hall.

"Mangroves are our lives," he said, as other villagers arrived, sitting on wooden benches to join our conversation. "Everything depends on the mangroves. Our houses are built of mangroves, the fish we catch live in the mangrove roots, the mangroves clean our air and we even get salt from the mangrove areas." Yet, he said gravely, the government had banned his people from harvesting mangroves—in the interests of conservation. It had also banned them from making clearings in the swamps to grow rice. It had even tried to ban fishing among the mangroves. This was crazy, he said. If they complied, they would starve. There was a chorus of agreement in the room.

The Rufiji delta is on nobody's radar. It was briefly famous during the First World War, when the most powerful German warship in the Indian Ocean, SMS *Königsberg*, was cornered there and sunk by British vessels. But, in its quiet way—and if the Tanzanian government only knew—its 130,000 acres of mangrove-covered swamps are an environmental success story. For centuries, the local people, known as Warufiji, have harvested the delta's resources without doing serious damage. They have fished the delta backwaters, cut mangroves and farmed rice. They have traded coconuts, prawns, mangrove poles and cashew nuts with Somalis, Persians, Omanis, Portuguese, Indians and, for a while, the British. Fleets of Arab dhows were once a common sight in the delta's waterways.

However, in the 1970s, the newly independent Tanzanian government made the foreign traders unwelcome. It wanted to promote national self-sufficiency; but it also said it wanted to protect the delta ecosystems from overexploitation. Since then, the Tanzanian Forest Service has fitfully tried to impose bans on exploitation of the delta's resources by its inhabitants. The Warufiji say this undermines their authority, reducing their ability to police the mangroves. It makes them outlaws in their own world. Their old system, under the watchful eye of village elders, worked well, they say. Their authority needs to be restored before outsiders come and wreck the place.

Foreign scientists who know the delta have tended to agree. The delta is no virgin forest, they say. It is already more used than conservationists might imagine from looking at its evident ecological health. That means it is also more resilient, says Betsy

Beymer-Farris of the University of Kentucky. Far from being on a roller-coaster ride to ecological decline, it is in a good place.

The Rufiji delta today is a backwater, economically and socially as well as ecologically. Current land use, says Beymer-Farris, "is not nearly as extensive as it was during the eighteenth and nineteenth centuries."[1] Most buildings are made of mud or wood, with thatch or tin roofs and no glass in their windows. Villages get their water from open wells. Most have no electricity other than from the solar cells that a few have erected on their roofs. There are hardly any roads and the dirt tracks are often impassable in the wet season. Travel is by boat. While some villages have a small clinic or a basic primary school, many have to travel for hours to reach either. No wonder literacy and life expectancy are low, even by Tanzanian standards. "Very few people get higher levels of education. And when they do, they rarely come back," one village chief told me.

Fitful government efforts to bring economic development to the delta have not been successful. Several villages have rusting signs left behind by long-forgotten and sometimes wildly misconceived projects. A plan to build the world's largest prawn farm would have privatized a third of the delta, but was abandoned after angry local protests.[2]

Nonetheless, the government seeks to control activities in the delta. It does this largely through its Forest Service, whose office on a road outside the delta was my first port of call. Its regional manager, Mathew Ntilicha, saw himself primarily as an environmental policeman. His task was to enforce a government ban on cutting mangroves and other uses of natural resources across an area of mangroves ten times the size of Manhattan. But he had only five staff and a single fiberglass boat to aid him. "We try to cooperate with communities, but sometimes we are fighting with them," he said. Outside his office sat a truck piled with confiscated timber. "The villagers do many illegal things. They believe the mangroves belong to them," one of his field staff said. "They remove the mangrove trees, and cultivate their rice in the mangrove areas. They run when they see us."

Villagers had their own stories of conflict. Such as when armed police torched three thousand huts as part of a forced evacuation of rice-farming areas in the north of the delta.[3] The good news was that Ntilicha wanted to change the relationship, to end the bans and introduce a land-use zoning system across the delta. "We want... to define areas for environmental protection, areas for cutting poles, areas for fishing, and areas for cultivation," he said. He wanted to "educate the villagers and give them responsibilities. There can be cutting and selling of trees, but in a proper way." Unsurprisingly, villagers were wary of losing control of their livelihoods, however much "cooperation" was on offer. They wanted to reassert their traditional methods of environmental control, which still hung on in the villages, despite lack of legal authority.

I was traveling with Ntilicha and Julie Mulonga of the NGO Wetlands International, which is offering expertise and trying to foster agreements between foresters and delta villages. Our boat took us through wide areas where mangroves lined the riverbanks and stretched across delta islands and peninsulas. Sediment brought down by the Rufiji River was forming mudbanks that were colonized by new mangrove saplings. Somewhere in the forests lurked wild pigs, monkeys, baboons, warthogs and mambas.[4] Elsewhere there was bare ground, where the Warufiji had cleared mangroves for rice paddies. Beside the paddies, abandoned straw huts sat charred, collapsing and falling into the water, victims of the government blitz on them a few years before.

At Mfisini, one of nineteen delta villages, the villagers described how their management of the mangroves had been thwarted by the Forest Service. "The village used to issue a license for people to cut the mangroves," village committee chair Salelie said. "We designated where and how many trees could be harvested." There were protected sacred groves and traditions and taboos about harvesting crops, he said. "It was all done in the village and it worked well."

Sitting next to him in his neat government-issue uniform, Ntilicha looked skeptical. The government had banned harvesting the mangroves because "the villagers had been taking things too far," he said. "There was no order." Even so, the ban was temporary, he insisted. "When we have a management plan, we will reinstate harvesting. But if you don't agree to the plan, the government will keep the ban." Salelie said he agreed with conservation, but wanted his village to regain its legal control. Clearly there was common ground. Equally clearly, it might all end in another round of arrests and hut burning.

Back in the Forest Service's boat, we headed on to Ruma, another small, well-organized village in the southern delta. It had an environment committee, and villagers were elected to manage activities such as mangrove planting, fishing and beekeeping. Those groups all said their authority was undermined by the laws imposed by the government. "It would be a good thing if we had charge of the licensing. The mangroves are our resource and we are the ones who would be affected if there are no trees," said Mohamed Hamis, the village environmental officer. He said their mangroves suffered from people cutting them for sale to outside traders. "The government cannot stop them," he said. But if his committee had authority, it could do the policing, because the villagers knew who the perpetrators were.

Our final stop was Jaja. For outsiders like us, coming from the land side of the delta, this coastal village was geographically the most remote, though it immediately felt more connected to the outside world. There was an old Arab dhow by the jetty—evidence that perhaps the old links with Arab traders across the Indian Ocean secretly persisted. There was an airstrip hidden in the bush too. The village had a soccer pitch and TV dishes. The forest sounds were interrupted by the dull thud of electronic music.

Unlike in some other delta villages, women played a leading role here. The dominant character at our village meeting was a woman. Dia Kiyonga arrived late and shook hands firmly with all the visitors. She was the queen of the mangroves, she said, listing the six local species and launching into an exhaustive description of their economic and ecological value—for house building, kindling fires, building beehives and fences, and providing poles for pounding grain. Then there were more specialized uses, including medicines, dyes and tools such as fishing floats, not to mention alcohol from fermented sap. Their roots, she said, were where fish bred and fed.

Because the Forest Service people didn't make it this far very often, the villagers said they had more control of their environment and suffered less interference. Far from losing their mangroves, Kiyonga said, Jaja had more than ever, "because we are taking better care of them. We have an area where we plant them, where there is wet mud rather than sand." What did she think of the government's plan for the mangroves? The queen of the mangroves waved away the very idea. "We don't need a government plan," she said. "We have our own plan. I am in charge of it. I tell people where and when to harvest. Or not."

SOME 2,500 MILES TO THE WEST, in Ghana, George Ayisi was drenched in sweat and sawdust as he painstakingly cleaned the teeth of his chainsaw, checked his fellows were out of harm's way, took a deep breath and resumed dismembering the freshly felled trunk of a giant hardwood tree. It was hard work on a hot and sunny Saturday morning. The tree was a hundred feet high and as wide as he was tall. It had taken an hour to fell. Now, with the tree suspended over a saw pit, he stood on the trunk to cut it into fifteen-foot lengths. Then he sawed the first of them lengthwise to take out quarter segments. His companions from the village down in the valley below rolled each quarter segment onto the ground, where they took it in turns to cut planks from it.

It was precision work. Barely a sliver of timber remained behind when they downed tools, balanced the planks on their heads and picked their way down the hillside, through a cocoa farm, to the road. "The government says this is illegal," said Ayisi, spitting out sawdust, "but how can they tell us not to do this? This is our land; these are our trees." He had paid the farmer fifty dollars to cut the tree, he said.

This scene, around fifty miles from the capital, Accra, close to the village of Brakumans, was typical of what happens across the country—and much of tropical Africa. Ayisi's planks were later picked up from the roadside by a truck and taken to a large lumber market in the nearby hill town of Akim Oda. The market, which employs some six hundred people, is one of dozens across the country—all entirely open and all outside the law. The country has an estimated hundred thousand chainsaw millers. The income from selling timber and the products it makes supports perhaps a million people.

Later, I toured the Akim Oda lumber market with Kwame Attafuah, a timber dealer and local organizer for Ghana's Domestic Lumber Traders' Association (DOLTA). Dozens of sheds were full of timber. Some of it was already being made into furniture or doors. "The government says we destroy the forest and create deserts. But it is lies, told by the big milling companies, who have ministers and officials in their pockets," Attafuah said. "We supply almost all the timber used in Ghana. All the officials and ministers buy from us. But they still make us illegal."

Since 1998, all production, transport and trade in lumber that has been turned into planks using chainsaws has been illegal in Ghana. Only lumber milled in sawmills is legal. This distinction has no obvious technical merit that I could see. It seemed designed to outlaw people like Ayisi and his friends, the poor freelance village artisans of the industry. Not that it stopped them. The chainsaw men still, as Attafuah said, supply almost all the timber used in the country.

At the Sethoo Wood Works at Asamankese, one of dozens of roadside carpentry shops we passed that day, they were turning chainsawed planks into wardrobes and chairs. It was a family business. The boss was training up his sons. He smiled sweetly at the suggestion that their timber was illegally cut and milled, and that his children might be on the wrong side of the law.

The ban leaves, as the only legal loggers, big companies with the cash to invest in sawmills and purchase licenses to log forest concessions. This legal business is dominated by a handful of large companies, which sell most of their timber for export to Europe, the US and Asia. The two parallel industries—one legal, commercial and for export; the other illegal, local and for domestic needs—consume similar amounts of timber across the country. With the rate of loss of Ghana's primary forests rising faster than any other tropical country, the question is: Which is best for Ghana and its forests?

The official argument for outlawing the small guys has been that big loggers maximize both production and conservation, are easier to police and can operate according to strict rules on sustainability. Maybe that is sometimes true. But as I saw in my morning with Ayisi, the chainsaw millers, often called sawyers, are highly selective, efficient and environmentally friendly loggers. They drag no heavy equipment into the forest, require no roads and select single trees, agrees Marieke Wit, a consultant on forest governance.[5] Any waste wood is left in the forests, where it can nurture the ecosystem, rather than on the floor of a sawmill. It was the big operators who were ransacking the forests, with heavy equipment and little concern about waste or damage.

This left me wondering about the wisdom of the demands of most environmentalists for governments to clamp down on illegal logging. They see it as the quickest and least controversial way of reducing deforestation across Africa. Surely, we are all against illegality? Clearly, where big business is illegally wrecking forests, that is a problem. Often a very big problem, as we have seen. Even so, many forest researchers I spoke to argue that an unthinking approach to legality is wrong. It assumes the laws are fair and make environmental sense, but often neither is true. Those researchers say that the illegal sawyers in Ghana are both environmentally more sustainable and socially and economically more useful than the legal commercial logging companies. The economic benefits they produce "are distributed more widely within communities than those provided by conventional logging," says Wit.

"Artisanal loggers are often seen as the primary cause of deforestation," says Alphonse Maindo of Tropenbos, a Dutch-based NGO dedicated to improving forest governance. "But their environmental impact is low, while their social benefit is high." What sense does it make for environmentalists to campaign against them? Surely, the sawyers are part of the solution and not part of the problem. Like shifting cultivators, they are due a PR makeover. As Bronson Griscom of The Nature Conservancy, which makes the case for the power of "nature-based solutions" to climate change, put it in 2013, "Loggers themselves can be a critical ally in helping us maintain biodiversity and mitigate climate change in tropical forests."[6]

Few can doubt the flagrant nature and corrupting effect in Ghana of making a mainstay of the national economy illegal. The outlawing of the artisans allows lower officials and police to extort bribes in return for turning a blind eye. Sitting in his office in the provincial town of Asamankese, DOLTA's regional secretary, Patrick Agyei, calculated for me the bribes paid to police for the daily passage of the twenty trucks that regularly carried sawyers' lumber from his region to Accra. At a going rate of $750 per truckload, it worked out at just over $100,000 a week.

Ghana is not alone. Across Central and West Africa—in Cameroon, Ivory Coast, Liberia and elsewhere—small-scale logging by locals is a much bigger contributor to local economies and employment than large-scale enterprises. Yet the local sawyers have for many years been outlawed, harassed and marginalized, while the big guys get the forests. It has generally been part of a wider policy of preventing communities from managing their own local forests.

There is change afoot, however. Just as governments are learning that local management of community forests is the route to sustainable use of the forests, so they are learning that local sawyers can be agents of that local control.

The Democratic Republic of the Congo is a heavily forested country better known for civil wars than environmental probity. But since 2016, it has allowed forest communities to apply for control of as much as 120,000 acres of nearby forest each, which they can exploit according to approved management plans. By early 2020, sixty-five of these community forests had been established, covering three million acres. According to Maindo, who is based in the country, the project is working: "When communities are granted local concessions, the management of artisanal logging improves, because each concession has a management plan."

Ghana too now has a plan to bring the artisanal loggers within the law, provided that they agree to train to cut up their logs using mobile sawmills rather than chainsaws. What it hasn't yet done is identify forests where they would legally be allowed to operate. The government has to bite the bullet by reallocating forest concessions from commercial exporters to communities, and allowing local sawyers to take charge of their harvesting. From Guatemala to Nepal, and the Amazon to the Congo, the evidence is that this would work—for the benefit of both communities and their forests.

George Ayisi may have looked an unlikely prospective forest savior as he cleaned the teeth of his chainsaw that Saturday morning outside Brakumans, but he was.

IT WOULD BE EASY to tell bad-news stories about the iniquities of how forests are managed across Africa and about the flouting of communities' rights to their forests. But in recent years there has been progress towards both fairness and a more enlightened approach to forest conservation. It has been fitful at times; Africa is not an easy place to make anything work well. Nonetheless, progress can be seen, especially when it is rooted in local control.

Take Kenya. This East African nation was the home of 2004 Nobel Peace Prize laureate Wangari Maathai, whose Green Belt Movement inspired millions of poor rural women in Kenya to plant trees. Four years before she won the prize, I interviewed her at her headquarters on the outskirts of Nairobi.[7] By then, she had established six thousand community nurseries and planted thirty million trees across the country on farms, in gardens, at the roadside, outside schools and public buildings, and even in forests. She had also pushed for community management of Kenya's forests, and attacked the corruption that caused the loss of many of the country's natural forests and common land under President Daniel arap Moi, who ruled from 1978 to 2002. "We are telling people they must learn to manage their local resources," she told me.

Moi had tried to see her off. She was beaten up, suffered endless death threats and for a time had to go into hiding. But in the end, she saw him off. After his demise amid a welter of corruption allegations, she served as an environment minister in the succeeding government. That government turned the tables with the passage of a Forest Act that created democratically elected Community Forest Associations that gave local people control over their forests and sought to protect the five forest areas regarded as the nation's "water towers": the Aberdare Mountains, the Mau Forest, Mount Kenya, Mount Elgon and the Cherangani Hills.

The towers cover just two percent of the country, but they capture moisture from the winds blowing west off the Indian Ocean. Much of that moisture is later transpired, helping maintain rains further inland; but much is also released into rivers. The water towers supply more than three-quarters of Kenya's surface water.[8]

The forests of the water towers had been reduced by deforestation, much of it organized by corrupt politicians keen to provide farmland for allies and constituents. "It was a political game, organized by Moi and his family," I was told in 2015 by Emilio Mugo, the newly installed director of the Kenya Forest Service. "It was a nightmare. The forest was being destroyed." The scandal had come to a head in 2009, he said, when an official report found that a quarter of the Mau Forest, the country's largest, had been lost in the previous fifteen years. The deforestation had caused a decline in several major rivers that begin in the forest. Communities and businesses along the rivers were running out of water.

Mugo showed me an unpublished annex to the report, naming the culprits in the destruction of the Mau Forest, including senior politicians and a member of the inquiry task force itself. His agency had been implicated in much of this, he agreed. For many years, it had been a quasi-military organization, often staffed with former soldiers still wearing fatigues. It had regarded local communities as the enemy, using extreme force to remove Indigenous communities such as the Ogiek people, whose homelands were in the Mau Forest. Usually this was done in the name of forest conservation. Usually to the detriment of forest conservation. The Ogiek were convenient scapegoats.

Under Mugo, however, the service transformed its purpose. It was about fostering good relations with the communities living on the slopes of the water towers, including the Ogiek. It regarded the three hundred or so community forest associations as its allies in looking after the forests. Members of the associations were allowed to use the forests for grazing livestock, cutting firewood and setting up beehives, provided that they undertook not to set up homes in the forest or cultivate forest land. I saw some of these projects in the field. At Kimunye village near Mount Kenya, Sarah Karungari showed me her association's dozen or so beehives in a clearing on the edge of the forest. It was a modest project, but in the old days, forest rangers would have torn down the hives and prosecuted the villagers.

The newfound cooperation between communities and the former forest enforcers worked both ways. "People who used to be poachers and illegal loggers are now defending the forests," said Simon Gitau, the senior warden for Mount Kenya. "These days, my main focus is maintaining good relations with the people who live here. They are my eyes and ears."

Other government agencies, including the Kenya Wildlife Service, which is in charge of national parks and the country's great populations of elephants, giraffes and other megafauna, are adopting the same community-centered approach. I met its assistant director for mountain conservation, Aggrey Naumo, over breakfast in a tourist lodge in the Aberdare National Park, another water tower. As we sipped our orange juice, elephants were grazing less than a hundred feet from our window. He said most of these animals spent the majority of their time outside his parks, on long migrations through the country's rich, wooded, agricultural landscapes. Many, perhaps most, of the country's trees were on farms rather than in forests. Maintaining those trees was vital to maintaining the wildlife. So collaboration with the communities was the name of his game.

"Farming communities know their ecosystems better than outsiders," Naumo said. "Better than my managers. We have to work with farming communities if we want to protect ecosystems, including the forests." Nature was repaying the attention. Forest cover on the Aberdare Mountains, the source of four of Kenya's seven largest rivers, had increased by a fifth since 2005. Most of that increase had come from natural regeneration.

I make no great claims for what has been happening in Kenya. Later, Mugo was dumped out of office by a minister he had crossed. Soon afterwards, his successor approved the demolition of three hundred Ogiek homes in the Mau Forest and more owned by another group, the Sengwer, in the Cherangani Hills—both in the name of protecting the forests, and despite robust evidence that local communities are the best defenders of their forests against outsiders.[9] In the face of these human rights atrocities, which appeared to intensify during the COVID-19 lockdown in the country, the EU canceled funding for protection of the water towers in late 2020, while promising to return it if the situation improved.[10]

But, despite such setbacks, recent history in Kenya suggests a possible future for trees, forests and forest communities. A future that can work in ways that meet the needs and desires of communities, that sustain their rights and could ensure the forests' future.

In 2021, the United Nations launched the Decade on Ecosystem Restoration. It was the latest in a growing crescendo of calls for a concerted effort to bring nature back, for the benefit of humanity as much as the planet. Central to that must be a great restoration of our forests and wider landscapes containing trees. Trees still cover almost a third of our planet's land, and more than ninety percent of our forests remain dominated by native species. So there is much to build on. If, as I believe, natural regrowth has to be the basis for the renaissance of the world's trees, then the custodians of that process must be the people who live in, among and from them. The evidence is that Indigenous and other forest communities are the best conservers of threatened forests and will be the best nurturers of the new forests. They know them best and need them most.

# POSTSCRIPT

## *Back Home*

I LIVE IN LONDON, a city of nine million people that in 2019 declared itself the world's first National Park City.[1] More than a third of this large urban area is public green space, much of it tree covered. Some boroughs have better tree cover than famous English natural landscapes such as the Lake District and the Yorkshire Dales.[2] London has roughly one tree for every inhabitant, but Londoners have often had to fight for this.

Close to my home in South London is Wandsworth Common. A thousand years old, it was surrounded by fields rather than houses, and was home to cattle and scrubby woodland rather than urban joggers. Ownership mostly rested with the Spencers, one of England's great aristocratic families. As London spread in the mid-nineteenth century, the Spencers wanted to sell the common to speculators for housing. The locals were having none of it. In 1869, and again the following year, there were mass protests. One poster, addressed to the "inhabitants and working men of Wandsworth and Battersea," asked, "Will you allow bankrupt and speculative builders, land societies, beershop keepers, railway companies, tailors, gentlemen and Noble Lords to rob you and your children of their common rights and footpaths, and the liberty of walking on

God's earth, without a struggle?" The call went out to tear down fences being erected to enclose ever more of the common. The fences were downed. The Spencers soon gave up the fight.[3] The common people secured the common land and its trees for common use in perpetuity. It remains open to all.

The great Wandsworth insurrection is not entirely forgotten. Many of the owners of today's million-pound houses around the common would no doubt be appalled at the mass violence that secured their green space. But in 2021, on the 150th anniversary of the common's future being secured with the Wandsworth Common Act, a small blue plaque went up to commemorate the chief troublemaker behind the protests, John Buckmaster. Much of the common is today open grass, but it still has more than 3,500 trees. An area that in the mid-nineteenth century housed the world's largest telescope is now set aside for natural tree regrowth.[4] It has some fifty tree species. A few locals fear this "jungle," wondering, like children in a fairy tale, who might be hiding among the dense foliage. Actually, in my experience, nobody hides there. The Big Bad Wolf is long gone, though the Spencers still have pubs, streets and schools named after them.

As I walk on the common it feels safe. And cool. I can feel every morning how much cooler it is than the nearby streets. Its transpiring stomata are better than any air conditioner. Its foliage also filters out noise and air pollutants, making the air cleaner. I like to think that this cool, green and tree-lined space on my doorstep exemplifies many of the themes in this book. It also holds lessons for how even the most urban of us can appreciate trees in our daily lives. You don't have to be a Xingu tribesperson or a Nepalese craftswoman, a Warufiji or a Wapichan to feel the benefit.

We have all heard of the urban heat island effect—how concrete and asphalt hold on to heat and fry us, especially in summer and at night. A study in Hong Kong, one of the world's most densely packed cities, found that in areas where tree density is thirty percent, temperatures are reduced by one degree Celsius (almost 2°F).[5] Even trees along sidewalks make a difference to urban areas that extends for a block or more. New York City averages between one and three degrees Celsius (about 2–5°F) warmer than the surrounding countryside—and as much as twelve degrees Celsius (about 20°F) warmer some evenings. But trees reverse the effect. Central Park is a cool summer haven.

You don't need to be imbued with well-being philosophy to realize that trees are good for our health and clear the mind. The data are clear. Public trees create calmer, safer spaces where people feel more able to walk, run or exercise. No doubt as a result, researchers in California found that ten-percent more canopy cover in a neighborhood reduced obesity by up to nineteen percent, and brought smaller benefits for sufferers from type 2 diabetes, asthma and high blood pressure.[6] None of this has yet persuaded Americans to turn back the tide of urban tree loss. A study by the US Forest Service found that each year in the past decade urban America lost 175,000 acres of tree cover—equating to some thirty-six million trees.[7] Oklahoma, Washington, DC, Rhode Island, Oregon and Georgia did worst.

The UK government's Office for National Statistics found that, by reducing air pollutants such as particulates and nitrogen oxides in city streets, urban trees saved nineteen hundred lives across the country and reduced health costs from heart and lung disease annually by an estimated £1 billion.[8] The bark of trees also reduces noise, one of the great bugbears of urban living. The larch turns out to be the best noise absorber among thirteen common urban tree species studied by University College London.

Besides soaking up pollution and noise and cooling the air, trees put us in touch with nature. With trees come birdsong and small mammals. Squirrels charge up and down the trees in my street. It seems that even a passing acquaintance with nature makes us healthier. Studies tracking individuals over several years routinely find that greenness reduces early deaths. One for the World Health Organization, covering eight million people in North America, Europe, Australia and China, found that for every ten-percent increase in greenery within sixteen hundred feet of where people lived, there was a four-percent reduction in premature mortality.[9] Using those figures, researchers for the US government's Forest Service concluded that if a city like Philadelphia upped its tree cover from twenty to thirty percent—the city's target for 2025—that would prevent four hundred premature deaths each year.[10]

The Japanese call communing with forests *shinrin-yoku*, meaning "forest bathing." Cynics may point out that the term was invented by officials at the country's forestry ministry. They may scoff at the hucksters from California to Camden offering expensive *shinrin-yoku* therapies, which seems rather the antithesis of the original impulse. But Japanese scientists claim that good peer-reviewed research shows the original unbranded, unmonetized forest bathing can improve the function of the human immune system for as much as thirty days while reducing anxiety, depression and anger, and lowering blood pressure.[11] Qing Li, an immunologist at the Nippon Medical School in Tokyo and president of the Society of Forest Medicine in Japan, swears by the healing powers of his local Rikugi-en Gardens.

Wild species in our midst and tree cover overhead have both also been correlated with feelings of well-being in dozens of cities around the world. Urban jungles are good. We need trees for our souls. Or, in the medicalized terminology of today, it is good for our mental health. The Hong Kong study mentioned earlier found that thirty-percent tree density cools our minds as well as the air, with a one-third reduction in the risk of psychological stress.[12] In Australia, among almost fifty thousand urbanites, those living near green spaces that have thirty-percent or more tree canopy were almost a third less likely to suffer psychological stress than those who have to make do with ten-percent canopy.[13]

So, thirty-percent tree cover looks like a good target for cities to aim for. It is the average for European cities, though much more than most cities worldwide manage outside small green neighborhoods. Cities can do it if they try. Singapore has thirty percent. A spot check on American cities by MIT highlighted Tampa's claim to thirty-six-percent tree cover. The US Forest Service calculates a figure of thirty-nine percent in New York City.[14] Citizen scientists organized by Forest Research, which is part of the UK Forestry Commission, found that average tree cover in 180 urban areas in England was only sixteen percent.[15]

Much more could be achieved. One study found that among thousands of cities across the world, seventeen percent of their land was open grass or shrub-covered land suitable for reforestation. Added up, these lands covered an area as large as Tennessee.[16] But despite the global potential, Europe is the only continent showing an increase in urban trees in recent years.[17] Some of its architects don't think that is enough and are trying to push the bounds by making trees integral parts of their buildings. The Hundertwasser House in Vienna, built in 1986, incorporates two hundred trees on balconies and terraces. The exterior of the Bosco Verticale apartments in downtown Milan is so festooned with trees—about eight hundred in all—that it appears from the outside almost entirely green. As the name says, it is a vertical forest. Green roofs abound in many other European cities too.

And beyond the cities, where the treescape is richer and more varied, there is even more to discover. I love my local park and I am sure that forest bathing has a real therapeutic benefit, but I also yearn for a bit more darkness, a bit more Hansel and Gretel. Give me the squat, gnarled and somewhat fearsome yews holed up in Kingley Bottom, tucked away in the hills of the South Downs National Park, a few miles northwest of the cathedral spire in Chichester.

Yews are not your average tree. Their wood is the world's hardest coniferous timber, as hard as iron. Once used for mill cogs and axles, yews also provided the timber of choice for longbows in the Middle Ages. Britain's oldest surviving wooden object is a thirteen-thousand-year-old spear made of yew and uncovered on an Essex beach. Their berries are remarkable too. They are the most poisonous of any English tree and were reputedly used by Druids in religious ceremonies. Later, the Normans began planting yews in churchyards, where, according to seventeenth-century astrologist and botanist Robert Turner, the tree "attracts and imbibes putrefaction."

There are said to be more yews in Kingley Bottom than anywhere in Europe. Some there are a thousand or more years old. One has a girth of twenty feet. Kingley Bottom was one of the country's first designated national nature reserves, but it is today a largely forgotten spot. Dark, dank and deserted, the trees there are perhaps too gloomy to attract day trippers. But when I walk among them—enjoying their cool air even on the hottest day—I am reminded afresh of the wonder of trees, their endurance and their value to us and our planet. They will, I suspect, outlive us all.

# ACKNOWLEDGMENTS

THIS BOOK HAS BEEN a long time in the making. The public urgency to reforest the planet may be quite recent, as is much of the science surrounding the importance of forests for our planet's life-support systems. But I have been reporting for decades on the angst over deforestation, and the strengthening case for placing the forests in the hands of those who known them best—forest inhabitants. Many of those I have relied on for expertise, wisdom and anecdotes are named directly. As a journalist, it is my habit to name my sources. But some people deserve particular gratitude for helping me achieve my goals.

In Brazil, huge gratitude to Michael Coe, who took a long flight from Boston and two overnight bus journeys to be with me on the forest frontier in Tanguro, and to Divino Silvério for his hospitality there. Also to Carlos Nobre for making time and to his brother Antonio for making mischief, to José Marengo for his wisdom and to Cláudio Almeida for not being daunted by his boss, President Bolsonaro. Thanks also to Britaldo Soares-Filhoand and Raoni Rajao in Belo Horizonte, and the people of IPAM in Brasília. It was great also to meet Philip Fearnside in his subterranean office in Manaus, after many years reporting his indefatigable work. Gerard and Margi Moss were unstintingly helpful in both Brasília and Lausanne.

Iris Moebius set up my vertiginous visit to the Amazon Tall Tower Observatory. Meinrat Andreae, whose work I have always read with gratitude, met me at the top. In Potsdam, thanks to Wei Weng and Kirsten Thonicke; in Stockholm to Lan Wang-Erlandsson, who taught me much about the evolving science of flying rivers; and in Zurich, to Jaboury Ghazoul and the go-getting Thomas Crowther. After several attempts, I shot the breeze with Anastassia Makarieva in Vienna. In the Netherlands, I met Ruud van der Ent and finally caught up with the forest polymath Douglas Sheil. In Israel, thanks to Dan Yakir and Yakir Preisler at the Weizmann Institute of Science, and Jay Shofet at the Society for the Protection of Nature in Israel.

In Indonesia, Jack Rieley facilitated my visit to the Sebangau swamp forest, on commission for Bill O'Neill at the *Guardian*. WWF's Yumiko Uryu and APRIL's Neil Franklin both helped me in Riau. I traveled to Ecuador with WWF to write about their work in the Pastaza valley: Lou Jost was an unexpected highlight. In China, I am indebted to the Yellow River Conservancy, especially Xue Yungpeng, for my journey to the Loess Plateau. James Fraser shared his R&R with me in Monrovia.

In Britain, I interviewed Peter Bunyard in Cornwall, Dominick Spracklen, Cat Scott and colleagues in Leeds, Edward Mitchard in Edinburgh, Sunitha Pangala in Milton Keynes and Nadine Unger in Exeter. Other people whose research permeates this book in a number of chapters include Simon Lewis, Chris Reij, Robin Chazdon, Erle Ellis, Simon Counsell, Melissa Leach, Sam Lawson, Liz Alden Wily, Clark Erickson, William Denevan and the departed heretics and loner warriors Michael Mortimore, Conrad Gorinsky and Patrick Darling.

My relationship with the forests NGO Fern, on whose board I sit, has been an inspiration in many ways. It also facilitated some of my field visits, including to Ghana, Liberia, Slovakia, where Peter Sabo of WOLF guided me, and to Scotland to meet reforesters with Hugh Chalmers and Mandy Haggith. Thanks especially to Saskia Ozinga and Hannah Mowat at Fern, but also to their many insightful colleagues. My section on the community forests of Nepal includes some material gathered by Hanna Aho. Among fellow board members, David Kaimowitz in particular has been a towering figure. He first spurred me on to investigate non-carbon aspects of forests' impact on climate—the "cool forests" agenda.

I traveled to the Rufiji delta in Tanzania on assignment for Wetlands International, and reported on the journey in *Water Lands*. Thanks there to Jane Madgwick and Julie Mulonga. The Drax Group took me to its pellet plants in Louisiana. Scot Quaranda of the Dogwood Alliance offered an alternative view. I spent time with the Wapichan in Guyana to write *Where They Stand* for the Forest Peoples Programme, mostly in the company of Tom Griffiths. Some insights came from working on a report for Oxfam and others on global land-rights issues, *Common Ground*. I also thank Andy White of the Rights and Resources Initiative for commissions to write several of their annual reviews.

Several trips and other researches were facilitated by journalistic commissions from Roger Cohn at *Yale Environment 360*, notably to Sarawak to investigate the slaying of Bill Kayong. You may also recognize here some snatches of text that have appeared for *Yale* on the Maya community forests, Tanguro, Australian bushfires and the environmental benefits of Indigenous reserves.

Some material here on biotic winds was commissioned by *Science*; and some on flying rivers by *New Scientist*. Thanks to commissioning editors Eric Hand and Rowan Hooper. I toured the Paraguayan Chaco with the World Land Trust and Guyra Paraguay. Subsequent material appeared in both *New Scientist* and my book *The Land Grabbers*, which also featured my Riau trip. My encounter with Lou Jost first appeared in *Deep Jungle*, commissioned by TV producer Brian Leith; and my journey to Chernobyl turned up in my books *The New Wild* and *Fallout*. I first wrote about my journey to China's Loess Plateau in *When the Rivers Run Dry*, and about the restoration of Puerto Rico's forests and Ascension

Island's Green Mountain in *The New Wild*. I wrote about the cacao agroforests of Cameroon in *Confessions of an Eco-sinner* thanks to an invitation from Anne Moorehead of the International Institute for Tropical Agriculture, who also got me to Sungbo's Eredo.

I am indebted also to a few PR companies. The good people at Burness put me in touch with several sources. Simon Forrester on behalf of the Goldman Prize set up several interviews, notably with Leng Ouch. Patrick Orr arranged my visit to the Kenyan water towers.

# NOTES


#### INTRODUCTION

1. www.loujost.com.
2. Richard Mabey, *The Cabaret of Plants: Botany and the Imagination* (London: Profile, 2015), 42.
3. Sir Walter Raleigh, *The Discovery of Guiana*, Bartleby: www.bartleby.com/33/74.html.
4. Cited in Susanna Hecht and Alexander Cockburn, *The Fate of the Forest: Developers, Destroyers and Defenders of the Amazon* (London: Verso, 1989), 11.
5. Henry M. Stanley, *Through the Dark Continent*, vol. 2 (New York: Harper & Brothers, 1878), 272, 212.
6. Richard Spruce, *Notes of a Botanist on the Amazon and Andes*, vol. 2 (London: Macmillan, 1908), 424.
7. Cited in Hecht and Cockburn, *The Fate of the Forest*, 1.
8. Mike Dash, "For 40 Years, This Russian Family Was Cut Off From All Human Contact, Unaware of World War II," *Smithsonian Magazine* (2013): www.smithsonianmag.com/history/for-40-years-this-russian-family-was-cut-off-from-all-human-contact-unaware-of-world-war-ii-7354256/.
9. Edward Wilson, *The Diversity of Life* (Cambridge, MA: Harvard University Press, 1992), 15.
10. Alex Kasprak, "Are There More Trees on Earth Than Stars in the Milky Way?" Snopes, December 12, 2016: www.snopes.com/fact-check/trees-stars-milky-way/.
11. Ohio State University, "Trees Can Slow Climate Change, But at a Cost," EurekAlert! (2020): www.eurekalert.org/pub_releases/2020-12/osu-tch120120.php.

#### 1: TREES ARE COOL

1. Tia Ghose, "The World's Largest Organism Is Dying," *Live Science* (2017): www.livescience.com/61116-mule-deer-are-eating-pando.html; Mabey, *The Cabaret of Plants*, 62.
2. David Beerling, *Making Eden: How Plants Transformed a Barren Planet* (Oxford: Oxford University Press, 2019).
3. Suzanne Simard et al., "Net Transfer of Carbon Between Ectomycorrhizal Tree Species in the Field," *Nature* (1997): doi.org/10.1038/41557; Diane Toomey, "Exploring How and Why Trees 'Talk' to Each Other,"

*Yale Environment 360* (2016): https://e360.yale.edu/features/exploring_how_and_why_trees_talk_to_each_other; Ferris Jabr, "The Social Life of Trees," *New York Times Magazine*, February 12, 2020: www.nytimes.com/interactive/2020/12/02/magazine/tree-communication-mycorrhiza.html.

4. Fred Sack and Jin-Gui Chen, "Pores in Place," *Science* (2009): doi.org/10.1126/science.1169553.
5. Dominick Spracklen et al., "Observations of Increased Tropical Rainfall Preceded by Air Passage Over Forests," *Nature* (2012): doi.org/10.1038/nature11390.
6. David Wilkinson, "The Parable of Green Mountain," *Journal of Biogeography* (2003): doi.org/10.1046/j.0305-0270.2003.01010.x.
7. Ibid.
8. David Catling and Stedson Stroud, "The Greening of Green Mountain, Ascension Island," University of Washington (2012): http://faculty.washington.edu/dcatling/Catling2012_GreenMountainSubmitted.pdf.
9. Daniel Kevles, "A Fistful of Wishful Thinking," *Nature* (1999): doi.org/10.1038/45683.
10. Mark Andrich and Jörg Imberger, "The Effect of Land Clearing on Rainfall and Fresh Water Resources in Western Australia: A Multi-Functional Sustainability Analysis," *International Journal of Sustainable Development and World Ecology* (2013): doi.org/10.1080/13504509.2013.850752.
11. John Boland, "Rainfall Enhances Vegetation Growth but Does the Reverse Hold?" *Water* (2014): doi.org/10.3390/w6072127.
12. David Ellison et al., "Trees, Forests and Water: Cool Insights for a Hot World," *Global Environmental Change* (2017): doi.org/10.1016/j.gloenvcha.2017.01.002.
13. Clifton Sabajo et al., "Expansion of Oil Palm and Other Cash Crops Causes an Increase of the Land Surface Temperature in the Jambi Province in Indonesia," *Biogeosciences* (2017): doi.org/10.5194/bg-14-4619-2017.
14. Ibid.
15. Avery Cohn et al., "Forest Loss in Brazil Increases Maximum Temperatures Within 50 Km," *Environmental Research Letters* (2019): https://iopscience.iop.org/article/10.1088/1748-9326/ab31fb.
16. Kirsten Findell et al., "The Impact of Anthropogenic Land Use and Land Cover Change on Regional Climate Extremes," *Nature Communications* (2017): doi.org/10.1038/s41467-017-01038-w.
17. Kimberly Novick and Gabriel Katul, "The Duality of Reforestation Impacts on Surface and Air Temperature," *Journal of Geophysical Research Biosciences* (2020): doi.org/10.1029/2019JG005543.
18. Yinon Bar-On et al., "The Biomass Distribution on Earth," *Proceedings of the National Academy of Sciences* (2018): doi.org/10.1073/pnas.1711842115.
19. Lan Qie et al., "Long-Term Carbon Sink in Borneo's Forests Halted by Drought and Vulnerable to Edge Effects," *Nature Communications* (2017): doi.org/10.1038/s41467-017-01997-0.
20. Simon Lewis et al., "Increasing Carbon Storage in Intact African Tropical Forests," *Nature* (2009): doi.org/10.1038/nature07771.

21. Shilong Piao et al., "Characteristics, Drivers and Feedbacks of Global Greening," *Nature Reviews Earth and Environment* (2019): doi.org/10.1038/s43017-019-0001-x.
22. Roel Brienen and Emanuel Gloor, "Across the World, Trees Are Growing Faster, Dying Younger—and Will Soon Store Less Carbon, *The Conversation* (2020): https://theconversation.com/across-the-world-trees-are-growing-faster-dying-younger-and-will-soon-store-less-carbon-145785.
23. Yan Li et al., "Local Cooling and Warming Effects of Forests Based on Satellite Observations," *Nature Communications* (2015): doi.org/10.1038/ncomms7603.
24. Almut Arneth et al., "Climate Change and Land: Special Report," Intergovernmental Panel on Climate Change (2019): www.ipcc.ch/site/assets/uploads/2019/08/Fullreport.pdf.
25. Christopher Williams et al., "Climate Impacts of U.S. Forest Loss Span Net Warming to Net Cooling," *Science Advances* (2021): doi.org/10.1126/sciadv.aax8859.

**2: FLYING RIVERS**

1. Eneas Salati and Peter Vose, "Amazon Basin: A System in Equilibrium," *Science* (1984): doi.org/10.1126/science.225.4658.129.
2. Ricardo Zorzetto, "Um Rio Que Flui Pelo Ar," *Pesquisa* (2009): https://revistapesquisa.fapesp.br/en/2009/04/01/a-river-that-flows-through-the-air/.
3. Partha Dasgupta, "Interim Report," *Dasgupta Review: Independent Review on the Economics of Biodiversity*, HM Treasury (2020): www.gov.uk/government/publications/interim-report-the-dasgupta-review-independent-review-on-the-economics-of-biodiversity.
4. Hubert Savenije, "New Definitions for Moisture Recycling and the Relationship With Land-Use Changes in the Sahel," *Journal of Hydrology* (1995): doi.org/10.1016/0022-1694(94)02632-L.
5. Ruud van der Ent, "A New View on the Hydrological Cycle Over Continents," Delft University of Technology (2014): https://repository.tudelft.nl/islandora/object/uuid:0ab824ee-6956-4cc3-b530-3245ab4f32be?collection=research.
6. Patrick Keys, "Analyzing Precipitationsheds to Understand the Vulnerability of Rainfall Dependent Regions," *Biogeosciences* (2012): doi.org/10.5194/bg-9-733-2012; Patrick Keys et al., "Revealing Invisible Water: Moisture Recycling as an Ecosystem Service," *PLOS One* (2016): doi.org/10.1371/journal.pone.0151993.
7. Patrick Keys, "Megacity Precipitationsheds Reveal Tele-connected Water Security Challenges," *PLOS One* (2018): doi.org/10.1371/journal.pone.0194311.
8. Amrit Dhillon, "Chennai in Crisis as Authorities Blamed for Dire Water Shortage," *Guardian*, June 19, 2019: www.theguardian.com/world/2019/jun/19/chennai-in-crisis-water-shortage-with-authorities-blamed-india.
9. Supantha Paul et al., "Weakening of Indian Summer Monsoon

Rainfall Due to Changes in Land Use Land Cover," *Scientific Reports* (2016): doi.org/10.1038/srep32177.

10. Tomo'omi Kumagai et al., "Deforestation-Induced Reduction in Rainfall," *Hydrological Processes* (2013): doi.org/10.1002/hyp.10060.
11. Clive McAlpine et al., "Forest Loss and Borneo's Climate," *Environmental Research Letters* (2018): doi.org/10.1088/1748-9326/aaa4ff.
12. Wei Weng et al., "Aerial River Management by Smart Cross-Border Reforestation," *Land Use Policy* (2019): doi.org/10.1016/j.landusepol.2019.03.010.
13. Gil Yosef et al., "Large-Scale Semi-Arid Afforestation Can Enhance Precipitation and Carbon Sequestration Potential," *Scientific Reports* (2018): doi.org/10.1038/s41598-018-19265-6.
14. Martin Claussen et al., "Simulation of an Abrupt Change in Saharan Vegetation in the Mid-Holocene," *Geophysical Research Letters* (1999): doi.org/10.1029/1999GL900494.
15. Roni Avissar and David Werth, "Global Hydroclimatological Teleconnections Resulting From Tropical Deforestation," *Journal of Hydrometeorology* (2005): doi.org/10.1175/JHM406.1.
16. Ibid.
17. David Medvigy et al., "Simulated Changes in Northwest U.S. Climate in Response to Amazon Deforestation," *Journal of Climate* (2013): doi.org/10.1175/JCLI-D-12-00775.1.
18. Solomon Gebrehiwot et al., "The Nile Basin Waters and the West African Rainforest: Rethinking the Boundaries," *WiresWater* (2018): doi.org/10.1002/wat2.1317.

### 3: FORESTS' BREATH

1. Amazon Tall Tower Observatory: www.attoproject.org.
2. Meinrat Andreae et al., "The Amazon Tall Tower Observatory (ATTO): Overview of Pilot Measurements on Ecosystem Ecology, Meteorology, Trace Gases, and Aerosols," *Atmospheric Chemistry and Physics* (2015): www.atmos-chem-phys.net/15/10723/2015/acp-15-10723-2015-discussion.html.
3. Laura Naranjo, "Volatile Trees," NASA Earthdata (2011): https://earthdata.nasa.gov/learn/sensing-our-planet/volatile-trees.
4. Nadine Unger, "To Save the Planet, Don't Plant Trees," *New York Times*, September 19, 2014: www.nytimes.com/2014/09/20/opinion/to-save-the-planet-dont-plant-trees.html.
5. Nadine Unger, "Human Land-Use-Driven Reduction of Forest Volatiles Cools Global Climate," *Nature* (2014): doi.org/10.1038/nclimate2347.
6. Gabriel Popkin, "How Much Can Forests Fight Climate Change?," *Nature* (2019): doi.org/10.1038/d41586-019-00122-z.
7. Cat Scott et al., "Impact on Short-Lived Climate Forcers Increases Projected Warming Due to Deforestation," *Nature Communications* (2018): doi.org/10.1038/s41467-017-02412-4.
8. Sunitha Pangala, "Large Emissions From Floodplain Trees Close

the Amazon Methane Budget," *Nature* (2017): doi.org/10.1038/nature24639.

9. Francis Bushong, "Composition of Gas From Cottonwood Trees," *Transactions of the Kansas Academy of Science* (1907): https://archive.org/details/jstor-3624516.
10. Mariza Costa-Cabral and Silvana Susko Marcelini, "The Role of Forests in the Maintenance of Stream Flow Regimes and Ground Water Reserves," Agroicone (2015): www.inputbrasil.org/wp-content/uploads/2015/05/The_role_of_forests_in_the_maintenance_of_stream_flow_regimes_and_ground_water_reserves_summary_Agroicone.pdf.
11. Fred Pearce, "A Controversial Russian Theory Claims Forests Don't Just Make Rain—They Make Wind," *Science* (2020): doi.org/10.1126/science.abd3856.
12. Anastassia Makarieva and Victor Gorshkov, "Biotic Pump of Atmospheric Moisture as Driver of the Hydrological Cycle on Land," *Hydrology and Earth System Sciences* (2007): doi.org/10.5194/hess-11-1013-2007.
13. Anastassia Makarieva et al., "Where Do Winds Come From? A New Theory on How Water Vapor Condensation Influences Atmospheric Pressure and Dynamics," *Atmospheric Chemistry and Physics* (2013): doi.org/10.5194/acp-13-1039-2013.
14. Bjorn Stevens and Sandrine Bony, "What Are Climate Models Missing?," *Science* (2013): doi.org/10.1126/science.1237554.
15. Douglas Sheil, "Forests, Atmospheric Water and an Uncertain Future: The New Biology of the Global Water Cycle," *Forest Ecosystems* (2018): doi.org/10.1186/s40663-018-0138-y.
16. Peter Bunyard et al., "Further Experimental Evidence That Condensation Is a Major Cause of Airflow," *Dyna* (2019): doi.org/10.15446/dyna.v86n209.73288.

**4: IN TANGURO**

1. Mateo Mier y Terán Giménez Cacho, "The Political Ecology of Soybean Farming Systems in Mato Grosso, Brazil," University of Sussex (2013): http://sro.sussex.ac.uk/id/eprint/48263/1/Mier_y_Ter%C3%A1n_Gim%C3%A9nez_Cacho%2C_Mateo.pdf.
2. IPAM, "Tanguro Project—Report" (2018): https://ipam.org.br/bibliotecas/tanguro-project-report/.
3. Michael Coe et al., "Feedbacks Between Land Cover and Climate Changes in the Brazilian Amazon and Cerrado Biomes," American Geophysical Union (2016): https://ui.adsabs.harvard.edu/abs/2016AGUFMGC23I..08C/abstract.
4. Michael Coe et al., "The Forests of the Amazon and Cerrado Moderate Regional Climate and Are the Key to the Future," *Tropical Conservation Science* (2017): doi.org/10.1177/1940082917720671.
5. Divino Silvério et al., "Testing the Amazon Savannization Hypothesis: Fire Effects on Invasion of a Neotropical Forest by Native Cerrado and Exotic Pasture Grasses," *Philosophical Transactions of the Royal Society B: Biological Sciences* (2013): doi.org/10.1098/rstb.2012.0427.

6. Thomas Lovejoy and Carlos Nobre, "Amazon Tipping Point: Last Chance for Action," *Science* (2019): doi.org/10.1126/science.1237554.
7. University of Helsinki, "Large-Scale Commodity Farming Accelerating Climate Change in the Amazon," Phys.org (2021): https://phys.org/news/2021-02-large-scale-commodity-farming-climate-amazon.html.
8. Martin Sullivan et al., "Long-Term Thermal Sensitivity of Earth's Tropical Forests," *Science* (2020): doi.org/10.1126/science.aaw7578.
9. Raoni Rajão et al., "The Rotten Apples of Brazil's Agribusiness," *Science* (2020): doi.org/10.1126/science.aba6646.
10. Fern, "100 Days of Bolsonaro—Ending the EU's Role in the Assault on the Amazon" (2019): www.fern.org/news-resources/100-days-of-bolsonaro-ending-the-eus-role-in-the-assault-on-the-amazon-945/.
11. Stephen Eisenhammer, "'Day of Fire': Blazes Ignite Suspicion in Amazon Town," Reuters, September 11, 2019: https://uk.reuters.com/article/uk-brazil-environment-wildfire-investiga/day-of-fire-blazes-ignite-suspicion-in-amazon-town-idUKKCN1VW1N9.
12. Britaldo Silveira Soares-Filho and Raoni Rajão, "Traditional Conservation Strategies Still the Best Option," *Nature Sustainability* (2018): doi.org/10.1038/s41893-018-0179-9.
13. MercoPress, "Brazil's Agribusiness Joins the Campaign to Stop Rainforest Fires in the Amazon," September 7, 2019: https://en.mercopress.com/2019/09/07/brazil-s-agribusiness-joins-the-campaign-to-stop-rainforest-fires-in-the-amazon.
14. Agence France-Presse, "British Supermarkets Threaten Brazil Boycott Over Amazon Land Law," Barron's, May 20, 2020: www.barrons.com/news/british-supermarkets-threaten-brazil-boycott-over-amazon-land-law-01589948105.
15. Greenpeace, "Amazon Deforestation Highest for 12 Years, Official Data Confirms," November 30, 2020: www.greenpeace.org.uk/news/amazon-deforestation-highest-for-12-years-official-data-confirms/.

**5: FIRES IN THE FOREST**

1. Lisa Cox, "More Than a Third of NSW Rainforests Found to Have Been Hit by Australian Bushfires," *Guardian*, June 6, 2020: www.theguardian.com/australia-news/2020/jun/07/more-than-a-third-of-nsw-rainforests-found-to-have-been-hit-by-australian-bushfires.
2. Australian Government, Department of Industry, Science, Energy and Resources, "Estimating Greenhouse Gas Emissions From Bushfires in Australia's Temperate Forests: Focus on 2019–20, Technical Update" (2020): www.industry.gov.au/data-and-publications/estimating-greenhouse-gas-emissions-from-bushfires-in-australias-temperate-forests-focus-on-2019-20.
3. Chris Lucas et al., "Bushfire Weather in Southeast Australia: Recent Trends and Projected Climate Change Impacts," Climate Institute of Australia (2007): doi.org/10.25919/5e31c82ee0a4c.

4. Matt Holden, "I'd Like a Raving Inner-City Lunatic T-Shirt for Christmas, Please," *Sydney Morning Herald*, November 12, 2019: www.smh.com.au/politics/federal/i-d-like-a-raving-inner-city-lunatic-t-shirt-for-christmas-please-20191112-p539tg.html.
5. NASA Earth Observatory, "Fires Take a Toll on Australian Forests" (2019): https://earthobservatory.nasa.gov/images/145998/fires-take-a-toll-on-australian-forests.
6. John Pickrell, "As Fires Rage Across Australia, Fears Grow for Rare Species," *Science* (2019): doi.org/10.1126/science.aba6144.
7. Dale Nimmo, "Animal Response to a Bushfire Is Astounding," *The Conversation* (2020): https://theconversation.com/animal-response-to-a-bushfire-is-astounding-these-are-the-tricks-they-use-to-survive-129327.
8. Matthias Boer et al., "Unprecedented Burn Area of Australian Mega-Forest Fires," *Nature Climate Change* (2020): doi.org/10.1038/s41558-020-0716-1.
9. John Pickrell, "Massive Australian Blazes Will 'Reframe Our Understanding of Bushfire,'" *Science* (2019): doi.org/10.1126/science.aba6144.
10. Adam J.P. Smith et al., "Climate Change Increases the Risk of Wildfires," *ScienceBrief Review* (2020): https://sciencebrief.org/uploads/reviews/ScienceBrief_Review_WILDFIRES_Sep2020.pdf.
11. University of East Anglia, "Climate Change Increases the Risk of Wildfires Confirms New Review" (2020): www.uea.ac.uk/about/-/climate-change-increases-the-risk-of-wildfires-confirms-new-review.
12. Anthony Westerling et al., "Briefing: Climate and Wildfire in Western US Forests," US Forest Service (2014): www.fs.usda.gov/treesearch/pubs/46580.
13. Matt Jolly et al., "Climate-Induced Variations in Global Wildfire Danger From 1979 to 2013," *Nature Communications* (2015): doi.org/10.1038/ncomms8537; Niels Andela et al., "A Human-Driven Decline in Global Burned Area," *Science* (2017): doi.org/10.1126/science.aal4108.
14. Hal Bernton, "Northwest Will Get More and Bigger Wildfires With Climate Changes," *Seattle Times*, September 11, 2017: https://phys.org/news/2017-09-northwest-forests-bigger-wildfires-climate.html.
15. Fred Pearce, "Wild About Fire," *New Scientist* (2000): www.newscientist.com/article/mg16822644-800-wild-about-fire/.
16. Michelle C. Mack et al., "Carbon Loss From Boreal Forest Wildfires Offset by Increased Dominance of Deciduous Trees," *Science* (2021): doi.org/10.1126/science.abf3903.
17. Kari Marie Norgaard and Sara Worl, "What Western States Can Learn From Native American Wildfire Management Strategies," *The Conversation* (2019): https://theconversation.com/what-western-states-can-learn-from-native-american-wildfire-management-strategies-120731.
18. Delilah Friedler, "California's Wildfire Policy Totally Backfired. Native Communities Know How to Fix it," *Mother Jones* (2019): www.motherjones.com/environment/2019/11/californias-wildfire-controlled-prescribed-burns-native-americans/.

19. Indigenous Peoples Burning Network: www.conservationgateway.org/ConservationPractices/FireLandscapes/FireLearningNetwork/Documents/FactSheet_IPBN.pdf.

20. Crystal Kolden, "We're Not Doing Enough Prescribed Fire in the Western United States to Mitigate Wildfire Risk," *Fire* (2019): doi.org/10.3390/fire2020030.

21. Norgaard and Worl, "What Western States Can Learn."

22. Kolden, "We're Not Doing Enough Prescribed Fire."

23. James Wilson et al., "Dozens Missing in Oregon as Historic Fires Devastate Western US," *Guardian*, September 12, 2020: www.theguardian.com/world/2020/sep/11/oregon-fires-california-washington-deaths-wildfires.

24. Michael Goss et al., "Climate Change Is Increasing the Likelihood of Extreme Autumn Wildfire Conditions Across California," *Environmental Research Letters* (2020): doi.org/10.1088/1748-9326/ab83a7.

25. Partha Dasgupta, "Interim Report," *Dasgupta Review: Independent Review on the Economics of Biodiversity*, HM Treasury (2020): www.gov.uk/government/publications/interim-report-the-dasgupta-review-independent-review-on-the-economics-of-biodiversity.

26. Gifford Miller, "Disentangling the Impacts of Climate and Human Colonization on the Flora and Fauna of the Australian Arid Zone Over the Past 100 Ka Using Stable Isotopes in Avian Eggshell," *Quaternary Science Reviews* (2016): doi.org/10.1016/j.quascirev.2016.08.009.

**6: LOST WORLDS**

1. Gaspar de Carvajal, in *The Discovery of the Amazon According to the Account of Friar Gaspar de Carvajal and Other Documents*, ed. Harry Heaton (New York: American Geographical Society, 1934).

2. Charles Clement et al., "The Domestication of Amazonia Before European Conquest," *Proceedings of the Royal Society B: Biological Sciences* (2015): doi.org/10.1098/rspb.2015.0813.

3. Ann Gibbons, "New View of Early Amazonia," *Science* (1990): doi.org/10.1126/science.248.4962.1488.

4. Michael Heckenberger et al., "Amazonia 1492: Pristine Forest or Cultural Parkland?," *Science* (2003): doi.org/10.1126/science.1086112.

5. Clark Erickson, "Raised Fields as a Sustainable Agricultural System From Amazonia," University of Pennsylvania (1994): https://repository.upenn.edu/anthro_papers/14/.

6. Umberto Lombardo et al., "Early Holocene Crop Cultivation and Landscape Modification in Amazonia," *Nature* (2020): doi.org/10.1038/s41586-020-2162-7.

7. Clark Erickson, "Amazonia: The Historical Ecology of a Domesticated Landscape," in *Handbook of South American Archaeology*, ed. Helaine Silverman and William Isbell (New York: Springer, 2008): doi.org/10.1007/978-0-387-74907-5_11.

8. Hans ter Steege et al., "Hyperdominance in the Amazonian Tree Flora," *Science* (2013): doi.org/10.1126/science.1243092.
9. Clement et al., "The Domestication of Amazonia Before European Conquest."
10. William Denevan, "The Pristine Myth: The Landscape of the Americas in 1492," *Annals of the Association of American Geographers* (1992): doi.org/10.1111/j.1467-8306.1992.tb01965.x.
11. Ibid.
12. Charles Kay, "Native Burning in Western North America: Implications for Hardwood Forest Management," USDA Forest Service (2020): www.fs.usda.gov/treesearch/pubs/23072.
13. Charles Mann, "1491," *Atlantic* (2002): www.theatlantic.com/magazine/archive/2002/03/1491/302445/.
14. Patrick Roberts et al., "The Deep Human Prehistory of Global Tropical Forests and Its Relevance for Modern Conservation," *Nature Plants* (2017): doi.org/10.1038/nplants.2017.93.
15. Damian Evans et al., "A Comprehensive Archaeological Map of the World's Largest Preindustrial Settlement Complex at Angkor, Cambodia," *Proceedings of the National Academy of Sciences* (2007): doi.org/10.1073/pnas.0702525104.
16. Chris Hunt, "Holocene Landscape Intervention and Plant Food Production Strategies in Island and Mainland Southeast Asia," *Journal of Archaeological Science* (2014): doi.org/10.1016/j.jas.2013.12.011.
17. Orenwalaju Asisi and David Aremu, "New Lights on the Archaeology of Sungbo Eredo, South-Western Nigeria," Dig It: Journal of the Flinders Archaeological Society (2016): www.academia.edu/25255530/New_Lights_on_the_Archaeology_of_Sungbo_Eredo_South_Western_Nigeria.
18. Patrick Darling, "Sungbo's Eredo—Africa's Largest Single Monument," Bournemouth University: https://csweb.bournemouth.ac.uk/africanlegacy/sungbo_eredo.htm.
19. Germain Bayon et al., "Intensifying Weathering and Land Use in Iron Age Central Africa," *Science* (2012): doi.org/10.1126/science.1215400.
20. Richard Oslisly et al., "Climatic and Cultural Changes in the West Congo Basin Forests Over the Past 5000 Years," *Philosophical Transactions of the Royal Society B: Biological Sciences* (2013): doi.org/10.1098/rstb.2012.0304.
21. Charles Mann, "The Real Dirt on Rainforest Fertility," *Science* (2002): doi.org/10.1126/science.297.5583.920.
22. Andre Junqueira et al., "The Role of Amazonian Anthropogenic Soils in Shifting Cultivation: Learning From Farmers' Rationales," *Ecology and Society* (2016): doi.org/10.5751/ES-08140-210112.
23. James Fraser et al., "An Intergenerational Transmission of Sustainability? Ancestral Habitus and Food Production in a Traditional Agro-ecosystem of the Upper Guinea Forest, West Africa," *Global Environmental*

*Change* (2015): doi.org/10.1016/j.gloenvcha.2015.01.013.

24. Victoria Frausin et al., "'God Made the Soil, but We Made It Fertile': Gender, Knowledge, and Practice in the Formation and Use of African Dark Earths in Liberia and Sierra Leone," *Human Ecology* (2014): doi.org/10.1007/s10745-014-9686-0.
25. STEPS Centre, "Anthropogenic Dark Earths in Africa?," University of Sussex (2010): https://steps-centre.org/wp-content/uploads/ADEcase.pdf.
26. Douglas Sheil et al., "Do Anthropogenic Dark Earths Occur in the Interior of Borneo: Some Initial Observations From East Kalimantan," *Forests* (2012): doi.org/10.3390/f3020207.
27. Erle Ellis et al., "Used Planet: A Global History," *Proceedings of the National Academy of Sciences* (2015): doi.org/10.1073/pnas.1217241110.
28. Roberts et al., "The Deep Human Prehistory of Global Tropical Forests."

### 7: THE WOODCHOPPER'S BALL

1. William Hickling Prescott, *History of the Conquest of Mexico* (London: Routledge, 1893), bk. 4, chap. 1: www.google.ca/books/edition/History_of_the_Conquest_of_Mexico/LGOWLSF5MIWC.
2. Stephanie Pain, "A Toast to Tonic," *New Scientist* (2019): doi.org/10.1016/s0262-4079(19)32439-x.
3. Mark Twain, *The Turning-Point in My Life* (1906): www.online-literature.com/twain/1324/.
4. Adam Hochschild, *King Leopold's Ghost: A Story of Greed, Terror, and Heroism in Colonial Africa* (London: Macmillan, 1999), 64.
5. Greg Grandin, *Fordlandia: The Rise and Fall of Henry Ford's Forgotten Jungle City* (New York: Metropolitan Books, 2009).
6. Tarnue Johnson, *A Critical Examination of Firestone's Operations in Liberia* (Bloomington: AuthorHouse, 2010).
7. Vivien Gornitz, "A Survey of Anthropogenic Vegetation Changes in West Africa During the Last Century: Climatic Implications," *Climatic Change* (1985): doi.org/10.1007/BF00144172.
8. Peter Chapman, *Jungle Capitalists: A Story of Globalisation, Greed and Revolution* (Edinburgh: Canongate, 2007).
9. Fred Pearce, "Brazil's Resettlement of Farmers Has Driven Amazon Deforestation," *New Scientist* (2015): www.newscientist.com/article/dn27993.
10. Ashwin Ravikumar et al., "Are Smallholders Really to Blame?," Forests News (2016): https://forestsnews.cifor.org/44155/are-smallholders-really-to-blame-2?fnl=.
11. William Laurance, "Roads to Ruin," *New York Times*, April 13, 2015: www.nytimes.com/2015/04/13/opinion/roads-to-ruin.html.
12. William Laurance, "If You Can't Build Well, Then Build Nothing at All," *Nature* (2018): doi.org/10.1038/d41586-018-07348-3.
13. Lucas Ferrante and Philip Fearnside, "The Amazon's Road to Deforestation," *Science* (2020): doi.org/10.1126/science.abd6977.

14. Thais Vilela et al., "A Better Amazon Road Network for People and the Environment," *Proceedings of the National Academy of Sciences* (2020): doi.org/10.1073/pnas.1910853117.
15. Neil Carter et al., "Road Development in Asia: Assessing the Range-Wide Risks to Tigers," *Science Advances* (2020): doi.org/10.1126/sciadv.aaz9619.
16. "Planned Roads Would Be 'Dagger in the Heart' for Borneo's Forests and Wildlife," James Cook University (2019): www.sciencedaily.com/releases/2019/09/190918142027.htm.
17. Andy Coghlan, "Africa's Road-Building Frenzy Will Transform Continent," *New Scientist* (2014): www.newscientist.com/article/mg22129512-800.
18. Fritz Kleinschroth et al., "Road Expansion and Persistence in Forests of the Congo Basin," *Nature Sustainability* (2019): doi.org/10.1038/s41893-019-0310-6.

**8: LOGGED OUT**

1. Borneo Nature Foundation, "Sebangau Landscape," www.borneonaturefoundation.org/sebangau-landscape/.
2. Fred Pearce, "Bog Barons: Indonesia's Carbon Catastrophe," *New Scientist* (2007): www.newscientist.com/article/mg19626321-600.
3. Sam Lawson, "Consumer Goods and Deforestation," Forest Trends (2014): www.forest-trends.org/wp-content/uploads/imported/for168-consumer-goods-and-deforestation-letter-14-0916-hr-no-crops_web-pdf.pdf.
4. Mikaela Weisse and Liz Goldman, "The World Lost a Belgium-Sized Area of Primary Rainforests Last Year," Global Forest Watch (2019): https://blog.globalforestwatch.org/data-and-research/world-lost-belgium-sized-area-of-primary-rainforests-last-year.
5. Charles Barber, "25 Years of So-Called 'Sustainable Forest Management' in Sarawak," World Resources Institute (2015): www.wri.org/blog/2015/11/25-years-so-called-sustainable-forest-management-sarawak.
6. Global Witness, "Inside Malaysia's Shadow State" (2013): https://cdn.globalwitness.org/archive/files/library/inside-malaysias-shadow-state-briefing.pdf.
7. Fred Pearce, "A Tale of Two Longhouses: More Air Conditioning Than Wild Men in the Heart of Borneo," *New Scientist* (1994): www.newscientist.com/article/mg14419525-700.
8. Fred Pearce, "Are Sarawak's Forests Sustainable?," *New Scientist* (1994): www.newscientist.com/article/mg14419534-000.
9. Sarawak Report, "We Release the Land Grab Data" (2011): www.sarawakreport.org/2011/03/we-release-the-land-grab-data/.
10. Gary Adit and Russell Ting, "Main Suspect Behind Bill Kayong Murder Caught," *Borneo Post*, December 11, 2016: www.theborneopost.com/2016/12/13/main-suspect-behind-bill-kayong-murder-caught/.
11. "Death Sentence Stuns Murderer," *Borneo Post*, August 11, 2018: www.theborneopost.com/2018/08/11/death-sentence-stuns-murderer/.

**9: CONSUMING THE FORESTS**


1. Peter Beaumont, "How a Tyrant's 'Logs of War' Bring Terror to West Africa," *Observer*, May 27, 2001: www.theguardian.com/world/2001/may/27/theobserver.
2. Global Witness, "Taylor-Made: The Pivotal Role of Liberia's Forests and Flag of Convenience in Regional Conflict" (2001): https://cdn.globalwitness.org/archive/files/pdfs/taylormade2.pdf.
3. Liberia Past and Present, "The Kouwenhoven Trial (2005–2017) and the Saga of His Extradition 2017–Present": www.liberiapastandpresent.org/TaylorCharles/KouwenhovenGus.htm.
4. Burness, "Report Reveals Dramatic Decline in Illegal Logging in Tropical Forest Nations," EurekAlert! (2010): www.eurekalert.org/pub_releases/2010-07/bc-rrd070810.php.
5. Delegation of the European Union to Liberia, "Fight Against Illegal Logging: The European Union and Liberia Sign Accord to Ensure Legal Origin of Imported Wood Products to the EU" (2011): www.eeas.europa.eu/archives/delegations/liberia/press_corner/all_news/news/2011/20110728_en.htm.
6. The Goldman Environmental Prize, "Leng Ouch": www.goldmanprize.org/recipient/leng-ouch/.
7. US Treasury, "Treasury Sanctions Corruption and Material Support Networks" (2019): https://home.treasury.gov/news/press-releases/sm849.
8. Natalie Sauer, "The Fight for the World's Largest Forest," *Climate Home News* (2019): www.climatechangenews.com/2019/10/08/siberia-illegal-logging-feeds-chinas-factories-one-woman-fights-back/.
9. Huang Wenbin and Sun Xiufang, "Tropical Hardwood Flows in China: Case Studies of Rosewood and Okoumé," Forest Trends (2013): www.forest-trends.org/wp-content/uploads/imported/tropical-hardwood-flows-in-china-v12_12_3_2013-pdf.pdf.
10. "EIA Accuses Indonesia and China of World's Biggest Smuggling Racket," Edie Newsroom (2005): www.edie.net/news/0/EIA-accuses-Indonesia-and-China-of-worlds-biggest-smuggling-racket/9581/.
11. Indra van Gisbergen, "Forest Crimes in Antwerp," Fern (2019): www.fern.org/story-articles/forest-crimes-in-antwerp/.
12. EIA Raw Intelligence, "'One Cannot Survive Going by the Rules': How Colossal Overharvesting and Money Laundering Became the Standard in the Forests of Gabon" (2019): https://eia-global.org/blog-posts/20190522-raw-intelligence-wcts-blog.
13. EIA, "The Croatian Connection Exposed" (2020): https://eia-international.org/wp-content/uploads/EIA-report-The-Croatian-Connection-Exposed-spreads.pdf.
14. Sam Lawson, "Stolen Goods: The EU's Complicity in Illegal Tropical Deforestation," Fern (2015): www.fern.org/fileadmin/uploads/fern/Documents/Stolen%20Goods_EN_0.pdf.
15. EIA, "For 20 Years We've Campaigned to Get China to Ban the Use of Illegal Timber—Now

It's Happening!" (2020): https://eia-international.org/press-releases/for-20-years-weve-campaigned-to-get-china-to-ban-the-use-of-illegal-timber-now-its-happening/.

16. EIA, "Corrupt Indonesian Timber Company Boss Jailed for Five Years," EIA (2020): https://eia-international.org/news/corrupt-indonesian-timber-company-boss-jailed-for-five-years-and-fined-more-than-180000/.
17. "Scientific Basis of EU Climate Policy on Forests: Open Letter," September 25, 2017: https://drive.google.com/file/d/0B9HP_Rf4_eHtQUpyLVIZZE8zQwc/view.
18. Anna Stephenson and David MacKay, "Life Cycle Impacts of Biomass Electricity in 2020," Department of Energy and Climate Change (2014): https://assets.publishing.service.gov.uk/government/uploads/system/uploads/attachment_data/file/349024/BEAC_Report_290814.pdf.
19. William Moomaw et al., "Focus on the Role of Forests and Soils in Meeting Climate Change Mitigation Goals," *Environmental Research Letters* (2020): doi.org/10.1088/1748-9326/ab6b38.
20. Adam Colette, "Uncovering the Truth: Investigating the Destruction of Precious Wetland Forests," Dogwood Alliance (2015): www.dogwoodalliance.org/tag/urahaw-swamp/.
21. Ember, "Playing With Fire: An Assessment of Company Plans to Burn Biomass in EU Coal Power Stations" (2019): https://ember-climate.org/project/playing-with-fire/.
22. Susanna Twidale, "Britain's Drax to Pilot Carbon Capture With Mitsubishi Heavy Industries," Reuters (2020): uk.reuters.com/article/us-drax-carboncapture/britains-drax-to-pilot-carbon-capture-with-mitsubishi-heavy-industries-idUKKBN23U3GT.

**10: NO-MAN'S-LAND**

1. Aldo Benitez, "Investigation Reveals Illegal Logging in Paraguay's Vanishing Chaco," Mongabay (2018): https://news.mongabay.com/2018/12/investigation-reveals-illegal-cattle-ranching-in-paraguays-vanishing-chaco/.
2. Carlos Barros, "Brazilian Meat Industry Encroaches on Paraguayan Chaco," Repórter Brasil (2018): https://reporterbrasil.org.br/2018/07/brazilian-meat-industry-encroaches-on-paraguayan-chaco/.
3. "The Impact of EU Consumption on Deforestation: Comprehensive Analysis of the Impact of EU Consumption on Deforestation," European Commission (2013): https://ec.europa.eu/environment/forests/pdf/1.%20Report%20analysis%20of%20impact.pdf.
4. "Deforestation in the Chaco Spikes in the Wake of 'Illegal' Presidential Decree Stripping Back Environmental Safeguards," Earthsight (2018): www.earthsight.org.uk/news/idm/deforestation-chaco-spikes-wake-illegal-presidential-decree.
5. "Promised Land," *Economist*, August 11, 2011: www.economist.com/the-americas/2005/08/11/promised-land.
6. Business and Human Rights Resources Centre, "Survival Intl.

Says Carlos Casado (Part of Grupo San José) Is Destroying Indigenous Group's Lands & Calls on Investors to Take Action" (2013): www.business-humanrights.org/en/paraguay-survival-intl-says-carlos-casado-part-of-grupo-san-jos%C3%A9-is-destroying-indigenous-groups-lands-calls-on-investors-to-take-action.

7. John Vidal, "Natural History Museum Expedition 'Poses Genocide Threat' to Paraguay Tribes," *Guardian*, November 8, 2010: www.theguardian.com/world/2010/nov/08/natural-history-museum-paraguay-tribes.
8. Spain's News, "Five Companies Dominate Global Trade in Deforestation-Causing Goods" (2020): https://spainsnews.com/five-companies-dominate-global-trade-in-deforestation-causing-goods/.
9. Fred Pearce, "Unilever Plans to Double Its Turnover While Halving Its Environmental Impact," *Daily Telegraph*, July 22, 2013: www.telegraph.co.uk/news/earth/environment/10188164/Unilever-plans-to-double-its-turnover-while-halving-its-environmental-impact.html.
10. Global Canopy, "Not on Target: Companies Must Do More to Deliver the New York Declaration on Forests" (2019): www.globalcanopy.org/sites/default/files/documents/resources/NYDFsignatoriesfinal.pdf.
11. Friends of the Earth US, "Largest US Investors Undermine Efforts to Halt Rainforest Destruction, New Report Finds," BankTrack (2020): www.banktrack.org/article/largest_us_investors_undermine_efforts_to_halt_rainforest_destruction_new_report_finds.
12. Andre Vasconcelos et al., "Illegal Deforestation and Brazilian Soy Exports: The Case of Mato Grosso," Trase (2020): http://resources.trase.earth/documents/issuebriefs/TraseIssueBrief4_EN.pdf.

**11: TAKING STOCK**

1. James Fairhead and Melissa Leach, "False Forest History, Complicit Social Analysis: Rethinking Some West African Environmental Narratives," *World Development* (1995): doi.org/10.1016/0305-750x(95)00026-9; Kate de Selincourt, "Demon Farmers and Other Myths," *New Scientist* (1996): www.newscientist.com/article/mg15020274-300.
2. Fred Pearce, "Counting Africa's Trees for the Wood," *New Scientist* (1994): www.newscientist.com/article/mg14219291-000.
3. Alexandra Tyukavina et al., "Congo Basin Forest Loss Dominated by Increasing Smallholder Clearing," *Science Advances* (2018): doi.org/10.1126/sciadv.aat2993.
4. Edward Mitchard and Clara Fintrop, "Woody Encroachment and Forest Degradation in Sub-Saharan Africa's Woodlands and Savannas 1982–2006," *Philosophical Transactions of the Royal Society B: Biological Sciences* (2013): doi.org/10.1098/rstb.2012.0406.
5. Global Forest Watch: www.globalforestwatch.org.
6. Food and Agriculture Organization of the United Nations,

"Global Forest Resources Assessment 2015," second edition (2016): www.fao.org/3/a-i4793e.pdf.

7. Peter Holmgren, "Can We Trust Country-Level Data From Global Forest Assessments?," Forests News (2015): https://forestsnews.cifor.org/34669/can-we-trust-country-level-data-from-global-forest-assessments.
8. Jean-Sébastien Landry and Navin Ramankutty, "Carbon Cycling, Climate Regulation, and Disturbances in Canadian Forests: Scientific Principles for Management," *Land* (2015): doi.org/10.3390/land4010083.
9. Anthony Swift and Jennifer Skene, "IPCC Report: Time to Reconsider How We Use Forests," NRDC (2019): https://www.nrdc.org/experts/anthony-swift/ipcc-climate-report-time-rethink-how-we-use-forests.
10. Bill Devall, ed., *Clearcut: The Tragedy of Industrial Forestry* (San Francisco: Sierra Club Books, 1993).
11. Frederic Beaudry, "Deforestation in Canada," Treehugger (2019): www.treehugger.com/deforestation-in-canada-1203594.
12. Government of Canada, "Deforestation in Canada: Key Myths and Facts": www.nrcan.gc.ca/our-natural-resources/forests-forestry/wildland-fires-insects-disturban/deforestation-canada-key-myths-and-facts/13419.
13. Landry and Ramankutty, "Carbon Cycling, Climate Regulation, and Disturbances in Canadian Forests."
14. Beaudry, "Deforestation in Canada."
15. Sylvie Gauthier et al., "Boreal Forest Health and Global Change," *Science* (2015): doi.org/10.1126/science.aaa9092.
16. Joshua Axelrod, "Pandora's Box," NRDC (2018): www.nrdc.org/sites/default/files/pandoras-box-clearcutting-boreal-carbon-dioxide-emissions-ip.pdf.
17. Praveen Noojipady et al., "Forest Carbon Emissions From Cropland Expansion in the Brazilian Cerrado Biome," *Environmental Research Letters* (2017): doi.org/10.1088/1748-9326/aa5986.
18. Robin Chazdon et al., "When Is a Forest a Forest? Forest Concepts and Definitions in the Era of Forest and Landscape Restoration," *Ambio* (2016): doi.org/10.1007/s13280-016-0772-y.
19. "Porous Definition of Forest Leaves Experts in Dilemma," *Nation* (2009): https://nation.africa/kenya/news/porous-definition-of-forest-leaves-experts-in-dilemma--617838.
20. Jean-Francois Bastin et al., "The Extent of Forest in Dryland Biomes," *Science* (2017): doi.org/10.1126/science.aam6527.
21. University of Copenhagen, "Artificial Intelligence Reveals Hundreds of Millions of Trees in the Sahara" (2020): https://news.ku.dk/all_news/2020/10/artificial-intelligence-reveals-hundreds-of-millions-of-trees-in-the-sahara/.
22. Chazdon et al., "When Is a Forest a Forest."
23. Philip Curtis et al., "Classifying Drivers of Global Forest Loss," *Science* (2018): doi.org/10.1126/science.aau3445.
24. Xiao Peng-Song et al., "Global Land Change From 1982 to 2016," *Nature* (2018): doi.org/10.1038/s41586-018-0411-9.

25. Peter Potapov et al., "The Last Frontiers of Wilderness: Tracking Loss of Intact Forest Landscapes from 2000 to 2013," *Science Advances* (2017): doi.org/10.1126/sciadv.1600821.
26. John Fa et al., "Importance of Indigenous Peoples' Lands for the Conservation of Intact Forest Landscapes," *Frontiers in Ecology and the Environment* (2020): doi.org/10.1002/fee.2148.
27. Claude Martin, *On the Edge: The State and Fate of the World's Tropical Rainforests* (Vancouver: Greystone, 2015), 142.
28. Serge Wich, "Distribution and Conservation Status of the Orang-utan (*Pongo* spp.) on Borneo and Sumatra: How Many Remain?," *Oryx* (2008): doi.org/10.1017/s003060530800197x.
29. David Morgan et al., "African Apes Coexisting With Logging: Comparing Chimpanzee (*Pan troglodytes troglodytes*) and Gorilla (*Gorilla gorilla gorilla*) Resource Needs and Responses to Forestry Activities," *Biological Conservation* (2018): doi.org/10.1016/j.biocon.2017.10.026.
30. Karel Mokany et al., "Reconciling Global Priorities for Conserving Biodiversity Habitat," *Proceedings of the National Academy of Sciences* (2020): doi.org/10.1073/pnas.1918373117.
31. Jeremy Hance, "Scientists Point to Research Flaw That Has Likely Exaggerated the Impact of Logging in Tropical Forests,"Mongabay (2013): https://news.mongabay.com/2013/01/scientists-point-to-research-flaw-that-has-likely-exaggerated-the-impact-of-logging-in-tropical-forests/.
32. David Edwards, "Degraded Lands Worth Protecting: The Biological Importance of Southeast Asia's Repeatedly Logged Forests," *Proceedings of the Royal Society B: Biological Sciences* (2010): doi.org/10.1098/rspb.2010.1062.

**12: FROM "STUMPS AND ASHES"**

1. Bill O'Driscoll, "A Century Ago, Pennsylvania Stood Almost Entirely Stripped of Trees," *Pittsburgh City Paper*, August 19, 2015: www.pghcitypaper.com/pittsburgh/a-century-ago-pennsylvania-stood-almost-entirely-stripped-of-trees/Content?oid=1848219.
2. Marcus Schneck, "Research Looks Into 400 Years of Pennsylvania's Forests, Which Have Been 'Completely Transformed,'" *Penn Live* (2019): www.pennlive.com/life/2019/11/new-research-into-400-years-of-change-in-pennsylvanias-forests.html; "Forests of Pennsylvania, 2015," USDA (2016): www.fs.fed.us/nrs/pubs/ru/ru_fs92.pdf; O'Driscoll, "A Century Ago."
3. Dina Spector, "American Forests Look Nothing Like They Did 400 Years Ago," *Business Insider* (2013): www.businessinsider.com/northeastern-us-forest-transformation-2013-9.
4. Jonathan Thompson et al., "Four Centuries of Change in Northeastern United States Forests," *PLOS One* (2013): doi.org/10.1371/journal.pone.0072540.
5. David Foster et al., "New England's Forest Landscape Ecological Legacies and Conservation Patterns Shaped by Agrarian History,"

Harvard Forest (2008): https://harvardforest.fas.harvard.edu/sites/harvardforest.fas.harvard.edu/files/publications/pdfs/Foster_NE_Forest_Landscape_2008.pdf.

6. David Foster et al., "Wildlands and Woodlands," Harvard Forest (2017): https://wildlandsandwoodlands.org/sites/default/files/Wildlands%20and%20Woodlands%202017%20Report.pdf.
7. *The Complete Poetical Works of Henry Wadsworth Longfellow*: www.gutenberg.org/files/1365/1365.txt.
8. Lee Klinger, "Ecological Evidence of Large-Scale Silviculture by California Indians," in *Unlearning the Language of Conquest*, ed. Wahinkpe Topa and Aka Don Trent Jacobs (Austin: Texas University Press, 2006): www.researchgate.net/publication/282780008_Ecological_evidence_of_large-scale_silviculture_by_California_Indians.
9. Gordon Bonan, "Forest, Climate and Public Policy: A 500-Year Interdisciplinary Odyssey," *Annual Review of Ecology, Evolution, and Systematics* (2016): www.annualreviews.org/doi/abs/10.1146/annurev-ecolsys-121415-032359.
10. Douglas MacCleery, "American Forests: A History of Resilience and Recovery," Forest History Society (2011): https://foresthistory.org/issues/american-forests-history-resiliency-recovery/.
11. Chloe Sorvino, "A Billion Dollar Fortune From Timber and Fire," *Forbes* (2018): www.forbes.com/feature/archie-emmerson-timber-forest-fires-logging/.
12. Gabriel Popkin, "How Small Family Forests Can Help Meet the Climate Challenge," *Yale Environment360* (2020): https://e360.yale.edu/features/how-small-family-forests-can-help-meet-the-climate-challenge.
13. MacCleery, "American Forests."
14. "Planting Trees Is No Panacea for Climate Change," University of California, Santa Cruz (2020): www.sciencedaily.com/releases/2020/05/200507143008.htm.
15. NASA Land Cover/Land-Use Change Program, "Afforestation in the Midwestern United States": https://lcluc.umd.edu/hotspot/afforestation-midwestern-united-states.
16. Ralph Alig and Brett Butler, "Area Changes for Forest Cover Types in the United States, 1952–1997, With Projections for 2050," USDA (2004): www.fs.fed.us/pnw/pubs/pnw_gtr613.pdf.
17. Ibid.
18. Grant Domke et al., "Greenhouse Gas Emissions and Removals From Forest Land, Woodlands, and Urban Trees in the United States, 1990–2018," US Forest Service (2020): www.nrs.fs.fed.us/pubs/59852.

**13: THE STRANGE REGREENING OF EUROPE**

1. Fred Pearce, "The Strange Death of Europe's Trees," *New Scientist*, December 4, 1986, 41.
2. Lucie Kupkova et al., "Forest Cover and Disturbance Changes, and Their Driving Forces: A Case Study in the Ore Mountains, Czechia, Heavily Affected by Anthropogenic Acidic Pollution

in the Second Half of the 20th Century," *Environmental Research Letters* (2018): doi.org/10.1088/1748-9326/aadd2c.

3. Earthsight, "Flatpacked Forests: Ikea's Illegal Timber Problem" (2020): www.earthsight.org.uk/flatpackedforests-en.
4. EIA, "Stealing the Last Forest" (2015): https://eia-global.org/reports/st; EIA, "Saving Europe's Last Virgin Forests" (2015): https://eia-global.org/subinitiatives/romania.
5. Guido Ceccherini et al., "Abrupt Increase in Harvested Forest Area Over Europe After 2015," *Nature* (2020): doi.org/10.1038/s41586-020-2438-y.
6. Tobias Kuemmerle et al., "Cross-border Comparison of Post-Socialist Farmland Abandonment in the Carpathians," *Ecosystems* (2008): doi.org/10.1007/s10021-008-9146-z.
7. Alexander Prishchepov et al., "Effects of Institutional Changes on Land Use: Agricultural Land Abandonment During the Transition From State-Command to Market-Driven Economies in Post-Soviet Eastern Europe," *Environmental Research Letters* (2012): doi.org/10.1088/1748-9326/7/2/024021; Natalia Kolecka et al., "Understanding Farmland Abandonment in the Polish Carpathians," *Applied Geography* (2017): dora.lib4ri.ch/wsl/islandora/object/wsl:14215.
8. Ana Kosnikovskaya, "What Is Happening to 100 Million Hectares of Forests in Russia?," Greenpeace International (2019): www.greenpeace.org/international/story/20631/what-is-happening-to-100-million-hectares-of-forests-in-russia/.
9. Irina Kurganova, "Carbon Cost of Collective Farming Collapse in Russia," *Global Change Biology* (2013): doi.org/10.1111/gcb.12379.
10. Bo Huang et al., "Predominant Regional Biophysical Cooling From Recent Land Cover Changes in Europe," *Nature Communications* (2020): doi.org/10.1038/s41467-020-14890-0.
11. Emma van der Zanden et al., "Trade-Offs of European Agricultural Abandonment," *Land Use Policy* (2017): doi.org/10.1016/j.landusepol.2017.01.003.
12. Sivan Kartha and Kate Dooley, "The Risks of Relying on Tomorrow's 'Negative Emissions' to Guide Today's Mitigation Action," Stockholm Environment Institute (2016): https://mediamanager.sei.org/documents/Publications/Climate/SEI-WP-2016-08-Negative-emissions.pdf.
13. Kim Naudts et al., "Europe's Forest Management Did Not Mitigate Climate Warming," *Science* (2016): doi.org/10.1126/science.aad7270; "Watch How Europe Is Greener Now Than 100 Years Ago," *Washington Post*, April 12, 2014: www.washingtonpost.com/news/worldviews/wp/2014/12/04/watch-how-europe-is-greener-now-than-100-years-ago/.
14. Samuel Zipper et al., "Land Use Change Impacts on European Heat and Drought: Remote Land-Atmosphere Feedbacks Mitigated Locally by Shallow Groundwater," *Environmental Research Letters* (2019): doi.org/10.1088/1748-9326/ab0db3.

15. Huang et al., "Predominant Regional Biophysical Cooling."
16. Fern, "Sweden: Taking Forestry Back to the Future" (2020): www.fern.org/news-resources/sweden-taking-forestry-back-to-the-future-2140/.
17. Matt McGrath, "'Wrong Type of Trees' in Europe Increased Global Warming," *BBC News*, February 5, 2016: www.bbc.co.uk/news/science-environment-35496350; Naudts et al., "Europe's Forest Management."
18. "EU Biodiversity Strategy for 2030: Bringing Nature Back Into Our Lives," European Commission (2020): www.arc2020.eu/wp-content/uploads/2020/05/Biodiversity-Strategy_draft_200423_ARC2020.pdf.
19. Heiko Schumacher et al., "More Wilderness for Germany: Implementing an Important Objective of Germany's National Strategy on Biological Diversity," *Journal for Nature Conservation* (2018): doi.org/10.1016/j.jnc.2018.01.002.
20. Kevin McKenna, "Scotland Has the Most Inequitable Land Ownership in the West. Why?," *Guardian*, August 10, 2013: www.theguardian.com/uk-news/2013/aug/10/scotland-land-rights.
21. Reforesting Scotland: www.reforestingscotland.org.
22. Carrifran Wildwood: www.carrifran.org.uk/about.

**14: FOREST TRANSITION**

1. Kathleen Buckingham and Craig Hanson, "The Restoration Diagnostic Case Example: China Loess Plateau," World Resources Institute (2015): https://files.wri.org/s3fs-public/WRI_Restoration_Diagnostic_Case_Example_China.pdf.
2. X. M. Feng et al., "Revegetation in China's Loess Plateau Is Approaching Sustainable Water Resource Limits," *Nature Climate Change* (2016): doi.org/10.1038/nclimate3092.
3. Rui Li et al., "Ecosystem Rehabilitation on the Loess Plateau," in *Regional Water and Soil Assessment for Managing Sustainable Agriculture in China and Australia*, ed. Tim McVicar et al., China-Australia Centre for International Agricultural Research (2002): https://pdfs.semanticscholar.org/28d3/8ac92cf0966ab8142f34745bc4d9d8b28784.pdf.
4. Karen Holl and Pedro Brancalion, "Tree Planting Is Not a Simple Solution," *Science* (2020): doi.org/10.1126/science.aba8232.
5. Fangyuan Hua et al., "Tree Plantations Displacing Native Forests: The Nature and Drivers of Apparent Forest Recovery on Former Croplands in Southwestern China From 2000 to 2015," *Biological Conservation* (2018): doi.org/10.1016/j.biocon.2018.03.034.
6. Minghong Tan and Xiubin Li, "Does the Green Great Wall Effectively Decrease Dust Storm Intensity in China? A Study Based on NOAA NDVI and Weather Station Data," *Land Use Policy* (2015): doi.org/10.1016/j.landusepol.2014.10.017.
7. Jon Luoma, "China's Reforestation Programs: Big Success or Just an Illusion?," *Yale Environment 360* (2012): https://e360.yale.edu/features/chinas_reforestation_programs_big_success_or_just_an_illusion.

8. Shixiong Cao et al., "Excessive Reliance on Afforestation in China's Arid and Semi-Arid Regions: Lessons in Ecological Restoration," *Earth-Science Reviews* (2011): doi.org/10.1016/j.earscirev.2010.11.002.
9. Fred Pearce, "Great Wall of Trees Keeps China's Deserts at Bay," *New Scientist* (2014): www.newscientist.com/article/mg22429994-900.
10. Ibid.
11. "Accounting Reveals That Costa Rica's Forest Wealth Is Greater Than Expected," World Bank (2016): www.worldbank.org/en/news/feature/2016/05/31/accounting-reveals-that-costa-ricas-forest-wealth-is-greater-than-expected.
12. Leighton Reid et al., "The Ephemerality of Secondary Forests in Southern Costa Rica," *Conservation Letters* (2018): doi.org/10.1111/conl.12607.
13. Pekka Kauppi et al., "Forest Resources of Nations in Relation to Human Well-Being," *PLOS One* (2018): doi.org/10.1371/journal.pone.0196248.
14. Kathleen Buckingham and Craig Hanson, "The Restoration Diagnostic Case Example: South Korea," World Resources Institute (2015): https://files.wri.org/s3fs-public/WRI_Restoration_Diagnostic_Case_Example_SouthKorea.pdf.
15. Kathleen Buckingham and Craig Hanson, "The Restoration Diagnostic Case Example: Panama Canal Watershed," World Resources Institute (2015): https://files.wri.org/s3fs-public/WRI_Restoration_Diagnostic_Case_Example_Panama.pdf.
16. Robert Hailmayr et al., "Impacts of Chilean Forest Subsidies on Forest Cover, Carbon and Biodiversity," *Nature Sustainability* (2020): doi.org/10.1038/s41893-020-0547-0.
17. Soumitra Ghosh and Larry Lohmann, "Compensating for Forest Loss or Advancing Forest Destruction?," World Rainforest Movement (2019): https://wrm.org.uy/wp-content/uploads/2019/09/WRM-Compensatory-Afforesation-in-India-2019.pdf.
18. European Commission, "The Impact of EU Consumption on Deforestation" (2019): http://ec.europa.eu/environment/forests/impact_deforestation.htm.

**15: TO PLANT OR NOT TO PLANT**

1. Amazing Forest, "Rebuilding the Amazon Rainforest One Tree at a Time," 24-7 Press Release (2011): www.24-7pressrelease.com/press-release/217869/rebuilding-the-amazon-rainforest-one-tree-at-a-time.
2. The Bonn Challenge: www.bonnchallenge.org.
3. John Vidal, "A Eureka Moment for the Planet: We're Finally Planting Trees Again," *Guardian*, February 13, 2018: www.theguardian.com/commentisfree/2018/feb/13/worlds-lost-forests-returning-trees.
4. NYDF Assessment Partners, "Protecting and Restoring Forests: A Story of Large Commitments Yet Limited Progress. Five-Year Assessment Report" (2019): https://forestdeclaration.org/images/uploads/resource/2019NYDFReport.pdf.

5. Simon Lewis et al., "Restoring Natural Forests Is the Best Way to Remove Atmospheric Carbon," *Nature* (2019): www.nature.com/articles/d41586-019-01026-8.
6. Simon Lewis and Charlotte Wheeler, "The Scandal of Calling Plantations 'Forest Restoration' Is Putting Climate Targets at Risk," *The Conversation* (2019): https://theconversation.com/the-scandal-of-calling-plantations-forest-restoration-is-putting-climate-targets-at-risk-114858.
7. Gregg Marland, "The Prospect of Solving the $CO_2$ Problem Through Global Reforestation," US Department of Energy (1988): https://books.google.co.uk/books/about/The_Prospect_of_Solving_the_CO2_Problem.html?id=RXQbNQAACAAJ.
8. Glen Peters, "The 'Best Available Science' to Inform 1.5°C Policy Choices," *Nature Climate Change* (2016): doi.org/10.1038/nclimate3000.
9. Bronson Griscom et al., "Natural Climate Solutions," *Proceedings of the National Academy of Sciences* (2017): doi.org/10.1073/pnas.1710465114.
10. Thomas Crowther et al., "Mapping Tree Density at a Global Scale," *Nature* (2015): doi.org/10.1038/nature14967.
11. Jean-Francois Bastin et al., "The Global Tree Restoration Potential," *Science* (2019): doi.org/10.1126/science.aax0848.
12. Nathaniel Hertz, "Trump Administration Will Eliminate Roadless Protections for Alaska's Tongass Forest," *Alaska Public Media* (2020): www.alaskapublic.org/2020/10/28/trump-administration-will-eliminate-roadless-protections-for-southeast-alaskas-tongass-forest/.
13. Alex Rudee, "3 Ways Biden and Congress Can Restore US Forests," World Resources Institute (2020): www.wri.org/blog/2020/11/3-ways-president-biden-and-congress-can-restore-us-forests.
14. Aisling Irwin, "The Ecologist Who Wants to Map Everything," *Nature* (2019): doi.org/10.1038/d41586-019-02846-4.
15. Erle Ellis et al., "Planting Trees Won't Save the World," *New York Times,* December 2, 2020: www.nytimes.com/2020/02/12/opinion/trump-climate-change-trees.html.
16. Irwin, "The Ecologist Who Wants to Map Everything."
17. World Rainforest Movement, "Driving 'Carbon Neutral': Shell's Restoration and Conservation Project in Indonesia" (2020): https://wrm.org.uy/articles-from-the-wrm-bulletin/section1/driving-carbon-neutral-shells-restoration-and-conservation-project-in-indonesia/.
18. Nathalie Seddon et al., "Getting the Message Right on Nature-Based Solutions to Climate Change," *Global Change Biology* (2021): doi.org/10.1111/gcb.15513.
19. "Biosphere-Atmosphere Fluxes," Weizmann Institute of Science: www.weizmann.ac.il/EPS/Yakir/biosphere-atmosphere-fluxes.
20. Yakir Preisler et al., "Mortality Versus Survival in Drought-Affected Aleppo Pine Forest Depends on the Extent of Rock Cover and Soil Stoniness,"

*Functional Ecology* (2019): doi.org/10.1111/1365-2435.13302.

21. "Turning the Desert Green," KKL–JNF: www.kkl-jnf.org/forestry-and-ecology/afforestation-in-israel/turning-the-desert-green/.
22. "From 'Improving Landscape' to Conserving Landscape: The Need to Stop Afforestation in Sensitive Natural Ecosystems in Israel and Conserve Israel's Natural Landscapes," Society for the Protection of Nature in Israel (2019): https://natureisrael.org/cms_uploads/Publications/Afforestation_Ecology_Damage_SPNI_2019.pdf; Guy Rotem et al., "Ecological Effects of Afforestation in the Northern Negev," Society for the Protection of Nature in Israel (2014): www.teva.org.il/_uploads/dbsattachedfiles/forestration_northern_negevspni_eng_finalmay2014.pdf.
23. Brandon Keim, "Not Everything Needs to Be a Forest," *Anthropocene* (2019): https://anthropocenemagazine.org/2019/12/savannas-should-not-be-forests/.
24. Kate Parr and Caroline Lehmann, "When Tree Planting Actually Damages Ecosystems," *The Conversation* (2019): https://theconversation.com/when-tree-planting-actually-damages-ecosystems-120786.

**16: LET THEM GROW**

1. Yunxia Wang et al., "Upturn in Secondary Forest Clearing Buffers Primary Forest Loss in the Brazilian Amazon," *Nature Sustainability* (2020): doi.org/10.1038/s41893-019-0470-4; Yunxia Wang et al., "Mapping Tropical Disturbed Forests Using Multi-decadal 30 m Optical Satellite Imagery," *Remote Sensing of Environment* (2019): doi.org/10.1016/j.rse.2018.11.028.
2. Philip Curtis et al., "Classifying Drivers of Global Forest Loss," *Science* (2018): doi.org/10.1126/science.aau3445.
3. Richard Houghton et al., "A Role for Tropical Forests in Stabilizing Atmospheric $CO_2$," *Nature Climate Change* (2015): doi.org/10.1038/nclimate2869.
4. Thomas Rudel, "When Fields Revert to Forest: Development and Spontaneous Reforestation in Post-War Puerto Rico," *Professional Geographer* (2004): doi.org/10.1111/0033-0124.00233.
5. Fei Yuan et al., "Forestation in Puerto Rico, 1970s to Present," *Journal of Geography and Geology* (2017): doi.org/10.5539/jgg.v9n3p30.
6. Susanna Hecht et al., "Globalization, Forest Resurgence, and Environmental Politics in El Salvador," *World Development* (2004): doi.org/10.1016/j.worlddev.2005.09.005.
7. Doribel Herrador-Valencia et al., "Tropical Forest Recovery and Socio-economic Change in El Salvador: An Opportunity for the Introduction of New Approaches to Biodiversity Protection," *Applied Geography* (2011): doi.org/10.1016/j.apgeog.2010.05.012.
8. Camila Rezende et al., "Atlantic Forest Spontaneous Regeneration at Landscape Scale," *Biodiversity Conservation* (2015): doi.org/10.1007/s10531-015-0980-y.
9. Deborah Zabarenko, "Tropical Rainforests Are Regrowing. Now What?," Reuters (2009):

www.reuters.com/article/us-rainforests/tropical-rainforests-are-regrowing-now-what-idUSTRE50B5CY20090112.

10. Robin Chazdon et al., "The Potential for Species Conservation in Tropical Secondary Forests," *Conservation Biology* (2009): doi.org/10.1111/j.1523-1739.2009.01338.x; Robin Chazdon, "Carbon Sequestration Potential of Second-Growth Forest Regeneration in the Latin American Tropics," *Science Advances* (2016): doi.org/10.1126/sciadv.1501639.
11. Susan Cook-Patton et al., "Mapping Carbon Accumulation Potential From Global Natural Forest Regrowth," *Nature* (2020): doi.org/10.1038/s41586-020-2686-x.
12. Robin Chazdon, *Second Growth: The Promise of Tropical Forest Regeneration in an Age of Deforestation* (Chicago: Chicago University Press, 2014).
13. Danae Rozendaal et al., "Biodiversity Recovery of Neotropical Secondary Forests," *Science Advances* (2019): doi.org/10.1126/sciadv.aau3114.
14. Tim Radford, "Natural Forests Are Best at Storing Carbon," Climate News Network (2020): https://climatenewsnetwork.net/natural-forests-are-best-at-storing-carbon/.
15. Xiao-Peng Song et al., "Global Land Change From 1982 to 2016," *Nature* (2018): doi.org/10.1038/s41586-018-0411-9.
16. Patrick Barkham, "Restore UK Woodland by Letting Trees Plant Themselves, Says Report," *Guardian*, December 15, 2020: www.theguardian.com/environment/2020/dec/15/restore-uk-woodland-trees-report-rewilding-britain.
17. Renato Crouzeilles et al., "Ecological Restoration Success Is Higher for Natural Regeneration Than for Active Restoration in Tropical Forests," *Science Advances* (2017): doi.org/10.1126/sciadv.1701345.
18. "Mangrove Restoration: To Plant or Not to Plant?," Wetlands International, www.wetlands.org/publications/mangrove-restoration-to-plant-or-not-to-plant/.
19. Ariel Lugo, "Emerging Forests on Abandoned Land: Puerto Rico's New Forests," *Forest Ecology and Management* (2003): doi.org/10.1016/j.foreco.2003.09.012.
20. Mary Duryea and Eliana Kampf, "Wind and Trees: Lesson Learned From Hurricanes," University of Florida (2007): https://hort.ifas.ufl.edu/woody/documents/FR173.pdf.
21. Maria Uriarte et al., "Hurricane María Tripled Stem Breaks and Doubled Tree Mortality Relative to Other Major Storms," *Nature Communications* (2019): doi.org/10.1038/s41467-019-09319-2.
22. Chazdon, *Second Growth*.

**17: AGROFORESTS**

1. Deborah Goffner et al., "The Great Green Wall for the Sahara and the Sahel Initiative as an Opportunity to Enhance Resilience in Sahelian Landscapes and Livelihoods," *Regional Environmental Change* (2019): doi.org/10.1007/s10113-019-01481-z.
2. UN Convention to Combat Desertification, "World Leaders Renew Commitment to Strengthen Climate Resilience Through Africa's Great Green Wall" (2015): www.unccd.int/news-events/world-leaders-renew-commitment-

strengthen-climate-resilience-through-africas-great.

3. Food and Agriculture Organization of the United Nations, "FAO Director-General Calls for Urgent Scale-Up of Africa's Great Green Wall," (2021): www.fao.org/news/story/en/item/1368800/icode/.
4. "Get Africa's Great Green Wall Back on Track," *Nature* (2021): doi.org/10.1038/d41586-020-03080-z.
5. Fred Pearce, "Interview: Can't See the Desert for the Trees," *New Scientist* (2008): www.newscientist.com/article/mg19726491-700.
6. Chris Reij et al., "Agroenvironmental Transformation in the Sahel: Another Kind of 'Green Revolution.'" International Food Policy Research Institute (2009): https://core.ac.uk/download/pdf/6257709.pdf.
7. Charlie Pye-Smith, "The Quiet Revolution: How Niger's Farmers Are Re-greening the Croplands of the Sahel," World Agroforestry Centre (2013): http://www.worldagroforestry.org/publication/quiet-revolution-how-nigers-farmers-are-re-greening-croplands-sahel.
8. "Tony Rinaudo," Right Livelihood Foundation (2018): www.rightlivelihoodaward.org/laureates/tony-rinaudo/; World Vision, "The Forest Maker" (2020): www.worldvision.com.au/global-issues/work-we-do/poverty/forest-maker.
9. John Carey, "The Best Strategy for Using Trees to Improve Climate and Ecosystems? Go Natural," *Proceedings of the National Academy of Sciences* (2020): doi.org/10.1073/pnas.2000425117.
10. Michael Mortimore, Mary Tiffen and Francis Gichuki, *More People, Less Erosion: Environmental Recovery in Kenya* (Chichester: Wiley, 1993).
11. Peter Holmgren, "Not All African Land Is Being Degraded," *Ambio* (1994): www.researchgate.net/publication/282684048.
12. Robert Zomer et al., "Trees on Farms: An Update and Reanalysis of Agroforestry's Global Extent and Socio-ecological Characteristics," World Agroforestry Centre (2014): http://apps.worldagroforestry.org/sea/Publications/files/workingpaper/WP0182-14.pdf.
13. Katie Reytar et al., "Deforestation Threatens the Mekong, but New Trees Are Growing in Surprising Places," World Resources Institute (2019): www.wri.org/blog/2019/10/deforestation-threatens-mekong-new-trees-are-growing-surprising-places.
14. Patrick Jagoret et al., "Long-Term Dynamics of Cocoa Agroforests: A Case Study in Central Cameroon," *Agroforestry Systems* (2011): doi.org/10.1007/s10457-010-9368-x.
15. University of Reading, "Agroforestry Is "Win Win" for Bees and Crops, Study Shows," Phys.org (2020): https://phys.org/news/2020-06-agroforestry-bees-crops.html.
16. James Robson, "Local Approaches to Biodiversity Conservation: Lessons From Oaxaca, Southern Mexico," *International Journal of Sustainable Development* (2008): doi.org/10.1504/IJSD.2007.017647.
17. Claire Kremen and Adina Merenlender, "The Important Complementary Role of Working Landscapes for Protected Area Effectiveness," Breakthrough

Institute (2018): https://thebreakthrough.org/issues/conservation/the-important-complementary-role-of-working-landscapes-for-protected-area-effectiveness.

18. Arild Angelsen and David Kaimowitz, "Agricultural Technologies and Tropical Deforestation," CAB International (2001): www.cifor.org/publications/pdf_files/Books/CAngelsen0101E0.pdf.
19. Vincent Ricciardi et al., "How Much of the World's Food Do Smallholders Produce?," *Global Food Security* (2018): doi.org/10.1016/j.gfs.2018.05.002.
20. Fred Pearce, "Will Intensified Farming Save the Rainforests?," *New Scientist* (2011): www.newscientist.com/article/mg20927986-200.
21. Thomas Rudel et al., "Agricultural Intensification and Changes in Cultivated Areas, 1970–2005," *Proceedings of the National Academy of Sciences* (2009): doi.org/10.1073/pnas.0812540106.
22. Simon Lewis et al., "Restoring Natural Forests Is the Best Way to Remove Atmospheric Carbon," *Nature* (2019): doi.org/10.1038/d41586-019-01026-8.
23. May Muthuri, "Could Tree Regeneration Hold Out Hope for Africa's Vulnerable Smallholder Farmers?," Regreening Africa (2020): https://regreeningafrica.org/project-updates/could-tree-regeneration-hold-out-hope-for-africas-vulnerable-smallholder-farmers/.
24. Mulugeta Lemenih and Habtemariam Kassa, "Re-greening Ethiopia: History, Challenges and Lessons," *Forests* (2014): doi.org/10.3390/f5081896; Peter Cronkleton et al., "How Do Property Rights Reforms Provide Incentives for Forest Landscape Restoration? Comparing Evidence from Nepal, China and Ethiopia," *International Forestry Review* (2017): www.cifor.org/publications/pdf_files/articles/ACronkleton1701.pdf.
25. Ibid.
26. Equator Initiative, "Abrha Weatsbha Community, Ethiopia, Case Study" (2013): www.equatorinitiative.org/wp-content/uploads/2017/05/case_1370354707.pdf.
27. Charlie Pye-Smith, "Friendly Fire," *New Scientist* (1997): www.newscientist.com/article/mg15621083-200.
28. Lorin Hancock, "Forest Fires: The Good and the Bad," WWF: https://www.worldwildlife.org/stories/forest-fires-the-good-and-the-bad.
29. Richard Schiffman, "Lessons Learned From Centuries of Indigenous Forest Management," *Yale Environment 360* (2018): https://e360.yale.edu/features/lessons-learned-from-centuries-of-indigenous-forest-management.

**18: INDIGENOUS DEFENDERS**

1. Size of Wales: https://sizeofwales.org.uk/tag/wapichan/.
2. Wikipedia, "Dadanawa Ranch," https://en.wikipedia.org/wiki/Dadanawa_Ranch.
3. Fergus MacKay, "The Wapichan People and the Guyanese Government Agree Terms of Reference for Formal Land Talks," Forest Peoples Programme (2016): www.forestpeoples.org/index.php/

en/enewsletters/fpp-e-newsletter-august-2016/news/2016/07/wapichan-people-and-guyanese-government-agree.

4. John Vidal, "Conrad Gorinsky Obituary," *Guardian*, September 12, 2019: www.theguardian.com/science/2019/sep/12/conrad-gorinsky-obituary.
5. Matthew Hallett et al., "Impact of Low-Intensity Hunting on Game Species in and Around the Kanuku Mountains Protected Area, Guyana," *Frontiers in Ecology and Evolution* (2019): doi.org/10.3389/fevo.2019.00412.
6. James McDonald, "Indigenous Reserves and the Future of the Amazon," *JSTOR Daily*, December 3, 2018: https://daily.jstor.org/indigenous-reserves-and-the-future-of-the-amazon/; Barbara Zimmerman, "Rain Forest Warriors: How Indigenous Tribes Protect the Amazon," *National Geographic* (2013): www.nationalgeographic.com/news/2013/12/131222-amazon-kayapo-indigenous-tribes-deforestation-environment-climate-rain-forest/.
7. Kathryn Baragwanath and Elia Bayi, "Collective Property Rights Reduce Deforestation in the Brazilian Amazon," *Proceedings of the National Academy of Sciences* (2020): doi.org/10.1073/pnas.1917874117.
8. Fred Pearce, "Give Forests to Local People to Preserve Them," *New Scientist* (2014): www.newscientist.com/article/dn25943.
9. "Secure Land Rights Essential to Protect Biodiversity and Cultures Within Indigenous Lands," University of East Anglia (2020): www.eurekalert.org/pub_releases/2020-05/uoea-slr051320.php.
10. Allen Blackman et al., "Titling Indigenous Communities Protects Forests in the Peruvian Amazon," *Proceedings of the National Academy of Sciences* (2017): doi.org/10.1073/pnas.1603290114.
11. Christoph Nolte et al., "Governance Regime and Location Influence Avoided Deforestation Success of Protected Areas in the Brazilian Amazon," *Proceedings of the National Academy of Sciences* (2013): doi.org/10.1073/pnas.1214786110.
12. Elaine Lopes et al., "Mapping the Socio-ecology of Non-Timber Forest Products (NTFP) Extraction in the Brazilian Amazon: The Case of Açaí (*Euterpe precatoria* Mart) in Acre," *Landscape and Urban Planning* (2019): doi.org/10.1016/j.landurbplan.2018.08.025.
13. Awudu Salami Sulemana Yoda, "'We Have Cut Them All': Ghana Struggles to Protect Its Last Old-Growth Forests," Mongabay (2019): https://news.mongabay.com/2019/08/we-have-cut-them-all-ghana-struggles-to-protect-its-last-old-growth-forests.
14. "Biodiversity Highest on Indigenous-Managed Lands," University of British Columbia (2019): www.sciencedaily.com/releases/2019/07/190731102157.htm.
15. Donald Hughes and Subash Chandran, "Sacred Groves Around the Earth: An Overview," in *Conserving the Sacred for Biodiversity Management*, ed. P. S. Ramakrishnan et al. (New Delhi: Oxford University Press, 1998).
16. "New Initiative Will Conserve Sacred Sites Rich in Biodiversity,"

Environmental News Service (2006): www.ens-newswire.com/ens/mar2006/2006-03-19-01.asp.

17. Kai Schmidt-Soltau, "Conservation-Related Resettlement in Central Africa: Environmental and Social Risks," *Development and Change* (2003): doi.org/10.1111/1467-7660.00317.
18. Daniel Brockington and James Igoe, "Eviction for Conservation: A Global Overview," Conservation and Society (2006): www.jstor.org/stable/26396619.
19. Ines Ayari and Simon Counsell, "The Human Cost of Conservation in Republic of Congo," Rainforest Foundation (2017): www.rainforestfoundationuk.org/media.ashx/the-human-impact-of-conservation-republic-of-congo-2017-english.pdf.
20. World Conservation Society, "Nouabalé-Ndoki National Park": https://programs.wcs.org/congo/Wild-Places/Nouabale-Ndoki-National-Park.aspx.
21. John Vidal, "Armed Ecoguards Funded by WWF 'Beat up Congo Tribespeople,'" *Guardian*, February 7, 2020: www.theguardian.com/global-development/2020/feb/07/armed-ecoguards-funded-by-wwf-beat-up-congo-tribespeople.
22. John Nelson, "Letter: Forest Peoples' Rights," *New Scientist* (2016): https://fdocuments.us/document/fly-by-chaos.html.
23. Survival International, "Exposed: WWF Execs Knew They Were Funding Rights Abuses in Africa, but Kept Report Under Wraps" (2019): https://www.survivalinternational.org/news/12247.
24. "Atrocities Prompt US Authorities to Halt Funding to WWF, WCS, in Major Blow to Conservation Industry," Survival International (2020): www.survivalinternational.org/news/12475.
25. Douglas Sheil and Eric Meijaard, "Purity and Prejudice: Deluding Ourselves About Biodiversity Conservation," *BioTropica* (2010): doi.org/10.1111/j.1744-7429.2010.00687.x.
26. Richard Schiffman, "Lessons Learned From Centuries of Indigenous Forest Management," *Yale Environment360* (2018): https://e360.yale.edu/features/lessons-learned-from-centuries-of-indigenous-forest-management.
27. Carol Linnitt, "How Indigenous People Are Changing the Way Canada Thinks About Conservation," *The Narwhal* (2018): https://thenarwhal.ca/how-indigenous-peoples-are-changing-way-canada-thinks-about-conservation/.
28. Government of Canada, "How Much Forest Does Canada Have?": www.nrcan.gc.ca/our-natural-resources/forests-forestry/state-canadas-forests-report/how-much-forest-does-canada-have/17601.
29. Xaxli'p Community Forest: www.xcfc.ca.
30. Sibyl Diver, "Community Voices: The Making and Meaning of the Xaxli'p Community Forest," Xaxli'p Community Forest Association (2016): www.researchgate.net/publication/330545376_Community_Voices_The_Making_and_Meaning_of_the_Xaxli'p_Community_Forest.
31. Donald Waller and Nicholas Reo, "First Stewards: Ecological

Outcomes of Forest and Wildlife Stewardship by Indigenous People of Wisconsin, USA," *Ecology and Society* (2018): doi.org/10.5751/ES-09865-230145.

32. Christopher and Barbara Johnson, "Menominee Forest Keepers," *American Forests* (2012): www.americanforests.org/magazine/article/menominee-forest-keepers/.
33. Alexandra Tempus, "The People's Forest," *Orion* (2018): https://orionmagazine.org/article/the-peoples-forest/.
34. Johnson, "Menominee Forest Keepers."

**19: COMMUNITY FORESTS**

1. Hanna Aho, "Nepal Shows How Forest Restoration Can Help People, Biodiversity and the Climate," Fern (2018): www.fern.org/news-resources/nepal-shows-how-forest-restoration-can-help-people-biodiversity-and-the-climate-122/.
2. Peter Gill, "In Nepal, Out-Migration Is Helping Fuel a Forest Resurgence," *Yale Environment 360* (2019): https://e360.yale.edu/features/in-nepal-out-migration-is-helping-fuel-a-forest-resurgence.
3. Harini Nagendra, "Drivers of Reforestation in Human-Dominated Forests," *Proceedings of the National Academy of Sciences* (2007): doi.org/10.1073/pnas.0702319104.
4. Kiran Paudyal et al., "Ecosystem Services From Community-Based Forestry in Nepal: Realising Local and Global Benefits," *Land Use Policy* (2017): doi.org/10.1016/j.landusepol.2017.01.046.
5. Rabin Raj Niraula et al., "Measuring Impacts of Community Forestry Program Through Repeat Photography and Satellite Remote Sensing in the Dolakha District of Nepal," *Journal of Environmental Management* (2013): doi.org/10.1016/j.jenvman.2013.04.006.
6. Rabin Raj Niraula and Niroj Timalsina, "Changing Face of the Churia Range of Nepal: Land and Forest Cover in 1992 and 2014," Helvetas (2015): www.academia.edu/13650797/Changing_Face_of_the_Churia_Range_of_Nepal_Land_and_Forest_Cover_in_1992_and_2014.
7. Johan Oldekop et al., "Reductions in Deforestation and Poverty From Decentralized Forest Management in Nepal," *Nature Sustainability* (2019): doi.org/10.1038/s41893-019-0277-3; Joe Stafford, "New Research Shows Community Forest Management Reduces Both Deforestation and Poverty," Manchester University (2019): www.manchester.ac.uk/discover/news/new-research-shows-community-forest-management-reduces-both-deforestation-and-poverty/.
8. Garrett Hardin, "The Tragedy of the Commons," *Science* (1968): doi.org/10.1126/science.162.3859.1243.
9. Garrett Hardin, "Who Benefits? Who Pays?," from *Filters Against Folly* (New York: Viking Penguin, 1985): www.garretthardinsociety.org/articles/art_who_benefits_who_pays.html.
10. Elinor Ostrom et al., "Revisiting the Commons: Local Lessons, Global Challenges," *Science* (1999): doi.org/10.1126/science.284.5412.278; Elinor Ostrom,

*Governing the Commons* (Cambridge: Cambridge University Press, 1990): https://wtf.tw/ref/ostrom_1990.pdf.

11. Walter Block, "Review of Elinor Ostrom's *Governing the Commons*," *Libertarian Papers* (2011): https://docplayer.net/50415890-Review-of-ostrom-s-governing-the-commons.html.
12. "Obituary: Elinor Ostrom," *Economist*, June 30, 2012: www.economist.com/obituary/2012/06/30/elinor-ostrom.
13. David Bray, "Toward 'Post-REDD+ Landscapes': Mexico's Community Forest Enterprises Provide a Proven Pathway to Reduce Emissions From Deforestation and Forest Degradation," CIFOR (2010): https://www.cifor.org/publications/pdf_files/infobrief/3272-infobrief.pdf.
14. Barbara Pazos-Almada and David Bray, "Community-Based Land Sparing: Territorial Land-Use Zoning and Forest Management in the Sierra Norte of Oaxaca, Mexico," *Land Use Policy* (2018): doi.org/10.1016/j.landusepol.2018.06.056.
15. Caleb Stevens et al., "Securing Rights, Combating Climate Change," World Resources Institute (2014): www.wri.org/publication/securing-rights-combating-climate-change.
16. Fred Pearce, "Closing the Gap," Rights and Resources Initiative (2016): https://rightsandresources.org/wp-content/uploads/RRI-2016-Annual-Review.pdf.
17. Luciana Porter-Bolland et al., "Community Managed Forests and Forest Protected Areas: An Assessment of Their Conservation Effectiveness Across the Tropics," *Forest Ecology and Management* (2012): doi.org/10.1016/j.foreco.2011.05.034.
18. Stevens et al., "Securing Rights, Combating Climate Change."
19. Ashwini Chhatre and Arun Agrawal, "Trade-Offs and Synergies Between Carbon Storage and Livelihood Benefits From Forest Commons," *Proceedings of the National Academy of Sciences* (2009): doi.org/10.1073/pnas.0905308106.
20. "Tribes Have Right to Maintain Relationship With Land," *Hindu Business Line* (2013): www.thehindubusinessline.com/economy/tribes-have-right-to-maintain-relationship-with-land-sc/article23104022.ece.
21. Fabio Suzart de Albuquerque et al., "Supporting Underrepresented Forests in Mesoamerica," *Natureza & Conservação* (2015): doi.org/10.1016/j.ncon.2015.02.001.
22. Jennifer Devine et al., "Drug Trafficking, Cattle Ranching and Land Use and Land Cover Change in Guatemala's Maya Biosphere Reserve," *Land Use Policy* (2020): doi.org/10.1016/j.landusepol.2020.104578.
23. Ileana Gomez and Ernesto Mendez, "Association of Forest Communities of Petén, Guatemala," CIFOR (2005): www.cifor.org/publications/pdf_files/books/bciforo801.pdf.
24. John Nittler and Henry Tschinkel, "Community Forest Management in the Maya Biosphere Reserve: Protection Through Profits," unpublished report submitted to USAID and others (2005);

Benjamin Hodgdon et al., "Deforestation Trends in the Maya Biosphere Reserve, Guatemala," Rainforest Alliance (2015): www.rainforest-alliance.org/wp-content/uploads/2021/07/MBR-Deforestation-Trends-1.pdf.

25. Liza Grandia, "Trickster Ecology: Climate Change and Conservation Pluralism in Guatemala's Maya Lowlands," in *Church, Cosmovision and the Environment*, ed. Evan Berry and Robert Albro (London: Taylor & Francis, 2018): doi.org/10.4324/9781315103785.
26. "Carmelita Cooperative: History and Present": https://turismocooperativacarmelita.com/en/carmelita-cooperative/history-present/.
27. Maya Nut Institute: https://mayanutinstitute.org/.
28. Rainforest Alliance, "Community: The Secret of Stopping Deforestation in Guatemala" (2018): www.rainforest-alliance.org/articles/community-the-secret-to-stopping-deforestation-in-guatemala; "25-Year Extension Granted to Community Forest Concession in Guatemala," *Forestry Journal* (2019): www.forestryjournal.co.uk/news/18115779.25-year-extension-granted-community-forest-concession-guatemala/.
29. Jaye Renold and Laura Sauls, "Paying With Their Lives to Protect the Forest," The Years Project: https://theyearsproject.com/learn/news/paying-with-their-lives-to-protect-the-forest/.
30. Dave Hughell and Rebecca Butterfield, "Impact of FSC Certification on Deforestation and the Incidence of Wildfires in the Maya Biosphere Reserve," Rainforest Alliance (2008): http://dk.fsc.org/preview.impacts-of-fsc-in-the-guatemala-maya-biosphere-reserve.a-240.pdf.
31. Kendra McSweeney et al., "Why Do Narcos Invest in Rural Land?," *Journal of Latin American Geography* (2016): doi.org/10.1353/lag.2017.0019.

**20: AFRICAN LANDSCAPES**

1. Betsy Beymer-Farris, "Taming the Tiger: The Political Ecology of Prawn Production in Tanzania," University of Illinois at Urbana-Champaign (2011): https://core.ac.uk/download/pdf/4834267.pdf.
2. Environmental Justice Atlas, "Protest Against Commercial Shrimp Farming in Rufiji Delta, Tanzania": https://ejatlas.org/conflict/protest-against-commercial-shrimp-farming-in-rujifi-delhi-tanzania.
3. REDD-Monitor, "WWF Scandal (Part 3): Embezzlement and Evictions in Tanzania" (2012): https://redd-monitor.org/2012/05/09/wwf-scandal-part-3-corruption-and-evictions-in-tanzania/.
4. Emmanuel Mbiha and Ephraim Senkondo, "A Socio-economic Profile of the Rufiji Floodplain and Delta: Selection of Four Additional Pilot Villages," Rufiji Environmental Management Project (2001): http://coastalforests.tfcg.org/pubs/REMP%2007%20TR%206%20vol%202%20Selection%20of%20Four%20Additional%20Pilot%20Villages.pdf.
5. Marieke Wit et al., "Chainsaw Milling: Supplier to Local Markets," European Tropical Forest Research Network (2010): www.researchgate.net/

publication/304110668_Chainsaw_milling_supplier_to_local_markets.

6. Bronson Griscom et al., "Natural Climate Solutions," *Proceedings of the National Academy of Sciences* (2017): doi.org/10.1073/pnas.1710465114; Bronson Griscom, "How Green Is Your Chainsaw?," Cool Green Science (2013): https://blog.nature.org/science/2013/10/08/how-green-is-your-chainsaw/.
7. Fred Pearce, "No, My President," *New Scientist* (2000): www.newscientist.com/article/mg16722484-800.
8. Kenyan Water Towers Agency, "Water Towers": https://watertowers.go.ke/water-towers/.
9. Community Land Action Now, "Kenyan Communities Report Illegal Evictions During COVID-19" (2020): www.forestpeoples.org/sites/default/files/documents/2020.07.23%20CLAN%20Press%20Statement_FINAL_.pdf.
10. Chris Lang, "EU Scraps US$35 Million Conservation and Climate Change Programme in Kenya Over Forced Evictions," *REDD-Monitor* (2020): https://redd-monitor.org/2020/10/01/eu-scraps-us35-million-conservation-and-climate-change-programme-in-kenya-over-forced-evictions/.

## POSTSCRIPT

1. London National Park City, "Launch July 2019": www.nationalparkcity.london/launch.
2. Wesley Stephenson, "Gardens Help Towns and Cities Beat Countryside for Tree Cover," *BBC News*, October 17, 2020: www.bbc.co.uk/news/science-environment-54311593.
3. Dorian Gerhold, *Wandsworth Past* (London: Historical Publications, 1998), 114–16.
4. Friends of Wandsworth Common, "Ecology and Nature": www.wandsworthcommon.org/the-common-wwco/#ecology-and-nature.
5. Edward Ng et al., "A Study on the Cooling Effects of Greening in a High-Density City: An Experience From Hong Kong," *Building and Environment* (2012): doi.org/10.1016/j.buildenv.2011.07.014.
6. Jared Ulmer et al., "Multiple Health Benefits of Urban Tree Canopy: The Mounting Evidence for a Green Prescription," *Health and Place* (2016): doi.org/10.1016/j.healthplace.2016.08.011.
7. Marlene Cimons, "America's Most Populated Areas Lose 175,000 Acres of Tree Cover Every Year," *Popular Science* (2018): www.popsci.com/city-trees-are-disappearing/.
8. Laurence Jones et al., "Developing Estimates for the Valuation of Air Pollution Removal in Ecosystem Accounts: Final Report," Office for National Statistics (2017): http://nora.nerc.ac.uk/id/eprint/524081/7/N524081RE.pdf.
9. David Rojas-Rueda et al., "Green Spaces and Mortality," *Lancet Planetary Health* (2019):doi.org/10.1016/S2542-5196(19)30215-3.
10. Michelle Konda et al., "Health Impact Assessment of Philadelphia's 2025 Tree Canopy Cover Goals," *Lancet Planetary Health* (2020): doi.org/10.1016/S2542-5196(20)30058-9.

11. Qing Li, "Effect of Forest Bathing Trips on Human Immune Function," *Environmental Health and Preventive Medicine* (2010): doi.org/10.1007/s12199-008-0068-3.
12. Ng et al., "A Study on the Cooling Effects of Greening."
13. Thomas Astell-Burt and Xiaoqi Feng, "Association of Urban Green Space With Mental Health and General Health Among Adults in Australia," *JAMA Network* (2019): doi.org/10.1001/jamanetworkopen.2019.8209.
14. Oliver Bach, "Green Streets: Which City Has the Most Trees?," *Guardian*, November 5, 2019: www.theguardian.com/cities/2019/nov/05/green-streets-which-city-has-the-most-trees.
15. Forest Research, "I-Tree Eco Projects Completed": www.forestresearch.gov.uk/research/i-tree-eco/i-tree-eco-projects-completed/.
16. Hoong Chen Teo et al., "Global Urban Reforestation Can Be an Important Natural Climate Solution," *Environmental Research Letters* (2021): doi.org/10.1088/1748-9326/abe783.
17. David Nowak and Eric Greenfield, "The Increase of Impervious Cover and Decrease of Tree Cover Within Urban Areas Globally (2012–2017)," *Urban Forestry and Urban Greening* (2020): doi.org/10.1016/j.ufug.2020.126638.

# FURTHER READING

There is a lot of great literature on trees, forests and their inhabitants. Here are some books that illuminated my journeys.

Ashmole, Myrtle and Philip. *The Carrifran Wildwood Story* (Jedburgh: Borders Forest Trust, 2009).

Beerling, David. *Making Eden: How Plants Transformed a Barren Planet* (Oxford: Oxford University Press, 2019).

Brum, Eliane. *The Collector of Leftover Souls: Dispatches From Brazil* (London: Granta, 2019).

Caufield, Catherine. *In the Rainforest* (London: Heinemann, 1985).

Chapman, Peter. *Jungle Capitalists: A Story of Globalisation, Greed and Revolution* (Edinburgh: Canongate, 2007).

Chazdon, Robin. *Second Growth: The Promise of Tropical Forest Regeneration in an Age of Deforestation* (Chicago: Chicago University Press, 2014).

Conrad, Joseph. *Heart of Darkness* (Harmondsworth: Penguin, 1973).

Deakin, Roger. *Wildwood: A Journey Through Trees* (London: Hamish Hamilton, 2007).

Desmond, Ray. *Kew: The History of the Royal Botanic Gardens* (London: Harvill, 1995).

Devall, Bill (ed). *Clearcut: The Tragedy of Industrial Forestry* (San Francisco: Sierra Club Books, 1993).

Diamond, Jared. *Collapse: How Societies Choose to Fail or Survive* (New York: Allen Lane, 2005).

Dowie, Mark. *Conservation Refugees: The Hundred-Year Conflict Between Global Conservation and Native Peoples* (Cambridge, MA: MIT Press, 2009).

Drayton, Richard. *Nature's Government: Science, Imperial Britain, and the "Improvement" of the World* (New Haven: Yale University Press, 2000).

Fairhead, James, and Melissa Leach. *Reframing Deforestation: Global Analyses and Local Realities—Studies in West Africa* (London: Taylor & Francis, 1998).

Grandin, Greg. *Fordlandia: The Rise and Fall of Henry Ford's Forgotten Jungle City* (New York: Metropolitan Books, 2009).

Greene, Graham. *Journey Without Maps* (London: Vintage, 2006).

Grove, Richard. *Green Imperialism: Global Expansion, Tropical Island Edens and the Origins of Environmentalism 1600–1860* (Cambridge: Cambridge University Press, 1995).

Harrison, Paul. *The Greening of Africa: Breaking Through in the Battle for Land and Food* (London: Paladin, 1987).

Hecht, Susanna, and Alexander Cockburn. *The Fate of the Forest: Developers, Destroyers and Defenders of the Amazon* (London: Verso, 1989).

Hochschild, Adam. *King Leopold's Ghost: A Story of Greed, Terror, and Heroism in Colonial Africa* (London: Macmillan, 1999).

Jones, Lucy. *Losing Eden: Why Our Minds Need the Wild* (London: Allen Lane, 2020).

Juniper, Tony. *Rainforest: Dispatches From the Earth's Most Vital Frontlines* (London: Profile, 2018).

Li, Qing. *Shinrin-Yoku: The Art and Science of Forest Bathing* (London: Viking, 2018).

Lovelock, James. *The Ages of Gaia: A Biography of Our Living Earth* (Oxford: Oxford University Press, 1988).

Mabey, Richard. *The Cabaret of Plants: Botany and the Imagination* (London: Profile, 2015).

Marris, Emma. *Rambunctious Garden: Saving Nature in a Post-wild World* (London: Bloomsbury, 2011).

Martin, Claude. *On the Edge: The State and Fate of the World's Tropical Rainforests* (Vancouver: Greystone, 2015).

McNeill, John. *Something New Under the Sun: An Environmental History of the Twentieth Century* (London: Allen Lane, 2000).

Mendes, Chico. *Fight for the Forest* (London: Latin America Bureau, 1989).

Monbiot, George. *Amazon Watershed: The New Environmental Investigation* (London: Abacus, 1991).

Monbiot, George. *Feral: Searching for Enchantment on the Frontiers of Rewilding* (London: Allen Lane, 2013).

Mortimore, Michael, Mary Tiffen, and Francis Gichuki. *More People, Less Erosion: Environmental Recovery in Kenya* (Chichester: Wiley, 1993).

O'Hanlon, Redmond. *Congo Journey* (London: Penguin, 1996).

Ostrom, Elinor. *Governing the Commons: The Evolution of Institutions for Collective Action* (Cambridge: Cambridge University Press, 1990).

Payne, Robert. *The White Rajahs of Sarawak* (Oxford: Oxford University Press, 1960).

Rackham, Oliver. *The History of the Countryside* (London: J. M. Dent, 1986).

Shanahan, Mike. *Ladders to Heaven: How Fig Trees Shaped Our History, Fed Our Imaginations and Can Enrich Our Future* (London: Unbound, 2016).

Thomas, Chris. *Inheritors of the Earth: How Nature Is Thriving in an Age of Extinction* (London: Allen Lane, 2017).

Tree, Isabella. *Wilding: The Return of Nature to a British Farm* (London: Picador, 2018).

Tudge, Colin. *The Secret Life of Trees: How They Live and Why They Matter* (London: Allen Lane, 2005).

Vince, Gaia. *Adventures in the Anthropocene: A Journey to the Heart of the Planet We Made* (London: Chatto & Windus, 2014).

Weber, Thomas. *Hugging the Trees: The Story of the Chipko Movement* (New Delhi: Penguin, 1988).

Wilson, Edward. *Half-Earth: Our Planet's Fight for Life* (New York: Liveright, 2015).

You might also find relevant
material in three of my own books:

*Deep Jungle* (London: Eden Project, 2005)

*The Landgrabbers: The New Fight Over Who Owns the Earth* (London: Eden Project, 2012)

*The New Wild: Why Invasive Species Will Be Nature's Salvation* (London: Icon, 2015)

# INDEX

Abasse, Tougiani, 214
Aberdare Mountains (Kenya), 267, 269
Abreu, Kátia, 70
acai, 236
acid rain, 168–69
Adams, John, 164
aerosols, 46
Africa: agroforestry, 219–20; carbon capture, 27; chainsaw timber millers in Ghana, 262–65, 266; community-centered forest protection in Kenya, 267–70; community control of forests, 248; deforestation, 103–4, 153; flying rivers and, 37; forest conservation, 266–67; grasslands, 193, 200; loss of ancient history, 93; reforestation and regrowth, 144–47, 211–12; roads, 107–8; Rufiji delta mangroves in Tanzania, 257–62; sacred groves, 249; small-scale logging, 265–66; Sungbo's Eredo, 91–92; *terra preta*, 95–96. *See also* Congo basin; *specific countries*
African Forest Landscape Restoration Initiative, 220
agriculture: deforestation from commercial agriculture, 137–38; forest regrowth following decline of, 146–47, 165–66, 170–72, 204–5; shifting cultivation, 96, 146, 153, 221–23; small-scale farmers and trees, 144–46. *See also* agroforestry
agroforestry: Africa, 219–20; biodiversity and, 218; cacao, 216–17; Ethiopia, 220–21; as global phenomenon, 216; Kenya, 215–16; Niger, 210–11, 212–14; shifting cultivation and, 221–23; sparers vs. sharers debate, 218–19; value of, 217–18
Agyei, Patrick, 265
Ahmed, Abiy, 221
air pollution, 168–69, 273
Akamba people, 215
albedo, 27–29, 198
Algeria, 211
Almeida, Cláudio, 68–69
Amanor, Kojo, 95
Amazing Forest, 188
Amazon: Amazon Tall Tower Observatory, 43–46, 47; and Brazilian politics and land speculation, 47, 67–71; carbon and, 27, 65–66; cold Amazon paradox, 55; colonial impacts, 86, 98; cooling effects, 25, 60–61;

destruction narrative, 66–67; Ecuador's cloud forests and, 2; fires, 44, 45, 63, 68–69, 71, 75, 76; flying river, 30–34; Fordlândia, 102; forest regrowth, 201–2; Indigenous peoples, 234–37; methane from, 50–51; migrant settlements, 105; North America impacts, 40–41; oxygen and, 20; pre-Columbian domestication, 85–88; pristine myth of, 86, 88–89; roads, 105–6; rubber extraction, 99–100; savannization, 62–64, 64–66, 76; *terra preta*, 94
Amazon Environmental Research Institute (IPAM), 61
Amazon Tall Tower Observatory (ATTO), 43–46, 47
American Forests, 243
Andela, Niels, 76
Andreae, Meinrat, 45, 46, 47
Andrich, Mark, 24
Angkor Wat, 91
Appalachian Mountains, 161
Arana, Julio César, 100
Arbor Day Foundation, 187
Arctic, 28, 37, 76, 195
Ascension Island, 22–23
Asia. *See* Southeast Asia; *specific countries*
Asia Pulp and Paper (AP&P), 113, 141
aspen, 17
Association of Forest Communities of Petén (ACOFOP), 254
Assynt Foundation, 176
Atlantic Forest (Brazil), 205
Atlantic Forest Restoration Pact, 205
Attafuah, Kwame, 263
Attenborough, David, 3
Australia: climate change and, 73; deforestation data, 147; fires, 72–75, 79, 80–81; flying rivers and rainfall, 24–25, 37; human-caused aridity, 80–81; sacred groves, 249; urban health benefits from trees, 274
*Avatar* (movie), 19
Avissar, Roni, 40–41
*Ayahuasca*, 6
Ayisi, George, 262–63, 266
Ayoreo people, 139–40
Aztecs, 97–98

Ban Ki-moon, 140
Bantu people, 93
Baranov, Vladimir, 124
Bastin, Jean-François, 152, 192
Bayaka people, 238–39
Beaudry, Frederic, 149, 150
BECCS (bioenergy with carbon capture and storage), 134
beef industry. *See* cattle ranching
Beerling, David, 18
Begotti, Rodrigo, 236
Belarus, 171
Belcher, Claire, 18
Belt and Road Initiative, 183
Bens, Samuel, 104
Beymer-Farris, Betsy, 258–59
biodiversity: agroforestry and, 218; forest regrowth and, 155–56, 207–8; Indigenous peoples and, 88, 237; loss of, 186, 199–200; Wapichan traditional lands in Guyana, 233
Bioenergy Bardejov, 129
biomass-burning power plants: BECCS (bioenergy with carbon capture and storage), 134; deforestation in Europe and, 170; Drax power station in UK, 130–33, 134; global popularity of, 133–34; Slovakia, 127–29
biotic pump: cold Amazon paradox and, 55; proof for, 55–57; resistance to, 54–55; Russia's Forest Code

and, 57–58; theory development, 52–54. *See also* flying rivers
Blackman, Allen, 236
Boland, John, 24–25
Bolivia, 39, 87, 100, 135
Bolsonaro, Jair, 47, 68, 69, 70–71, 106
Bond, William, 200
Bonn Challenge, 188–89, 190, 191–92, 200, 220
Borders Forest Trust, 177
boreal forests, 75, 148–51, 241, 249
Borlaug, Norman, 218, 219
Borneo: biodiversity and forest regrowth, 155; carbon capture, 27; Dayak people, 117–20, 240; illegal logging, 109–11, 125; pre-colonial domestication, 91; rainfall decline and deforestation, 39; Sarawak deforestation, 115–20; *terra preta*, 95–96
Bosco Verticale apartments (Milan), 275
Bowman, David, 76
Brando, Paulo, 76
Brandt, Martin, 152
Bray, David, 247–48
Brazil: Amazonian politics and land speculation, 47, 67–71; Atlantic Forest, 205; Bonn Challenge and, 189; cattle ranching (beef industry), 67–68, 69–70, 202, 219; flying river, 30–34; intact forests in, 154; meeting forest restoration goals, 202; migrant settlements in Amazon, 105; rubber war with Bolivia, 100; soy industry, 60, 67, 68, 142, 219; TreeSisters and, 187. *See also* Amazon
Brazilian Confederation of Agriculture and Livestock, 70
British Columbia, 75, 149, 241–42
Brito, Marcello, 70
Brockington, Dan, 238
Brunei, 185
Buckmaster, John, 272
Buddhism, 5
Bunyard, Peter, 56–57
burial and memorial sites, 249
Burkina Faso, 220
Bushong, Francis, 51

cacao, 97, 101, 216–17, 249
Calderón, Rafael, 184
California, 40–41, 75, 77–78, 80, 164, 249
Cambodia, 91, 123–24
Cameron, James: *Avatar* (movie), 19
Cameroon, 95, 146, 187, 217, 239, 265
Canada: albedo and conifer forests, 28; contradictory data on deforestation, 147–51; fires, 75; flying rivers, 37; Indigenous land stewardship, 240–42; intact forests in, 154
Cao, Shixiong, 182
carbon and carbon emissions: accumulation in trees, 18, 26–27; Amazon and, 65–66; from Australian fires, 73; biomass-burning power plants and, 131–32, 134; from deforestation, 26; European forests and, 173; forest regrowth and, 205–6, 213; from global warming, 66; offset projects, 194–95; tree planting and, 190–95; US reforestation and, 167
Carpathian Mountains, 128, 170
Cartes, Horácio, 138
Carvajal, Gaspar de, 85
Casaccia, José Luis, 138
Casado, Carlos, 138–39
Casement, Roger, 101
Castro Salazar, René, 184
cattle ranching (beef industry): Brazil, 67–68, 69–70, 202, 219; Chaco region, Paraguay, 136–37, 140; community-managed in Switzerland, 246–47; Costa Rica,

183–84; Guatemala, 251, 255; intensification of and deforestation, 219; southern US, 166; Wapichan traditional land in Guyana, 229
Center for International Forestry Research (CIFOR), 248
Central African Republic, 239
Central America. *See* Latin America
Chaco region (Paraguay), 135–37, 138–40
Chad, 95
chainsaws: deforestation and, 104; small-scale logging within Ghana, 262–65, 266
Chalmers, Hugh, 177
Chaulagai, Ramhari, 244
Chazdon, Robin, 194, 206, 209, 218, 222
Cheng Hangjun, 180
Chennai, 38, 39
Chen Youping, 124
Cherangani Hills (Kenya), 267, 269
Chernobyl, 208–9
Cherubini, Francesco, 171
Chhatre, Ashwini, 250
chicle, 253
Chile, 185, 186
China: Belt and Road Initiative, 183; Bonn Challenge and, 189; Cambodian rosewood and, 123; deforestation data, 147; flying rivers and, 36, 38, 42, 52; Grain for Green program, 179, 181; Great Green Wall, 181–82; international timber market and, 124–26; Loess Plateau Watershed Rehabilitation Project, 179–81; reforestation, 10, 179–82, 185, 200; road building in Africa, 107; sacred groves, 249
chocolate. *See* cacao
Christianity, 5
cinchona, 98, 101
cities, 25, 38, 272–75
Clement, Charles, 86, 88–89, 94
climate change: albedo and, 27–29; Australia and, 73; fires and, 75–76, 80; global greening and carbon capture by trees, 26–27; natural regeneration and, 206–7; Paris Agreement, 191, 220; savannization and, 62–64, 64–66, 76, 80; volatile organic compounds and, 48–50
cloud condensation nuclei, 46–47
cloud forests (Ecuador), 1–4, 34
clouds, 46, 48. *See also* flying rivers
coca (cocaine), 98
Coe, Michael, 61, 62, 63, 66
cold war, 169–70
colonialism: Amazon, 86, 98; Australia, 81; Congo, 100–101; deforestation from, 103–4, 164–65; fires and, 78; Guyana, 229; industrial exploitation by, 101–3; migrant resettlement in forests, 105; North America, 90; pillaging South American resources, 97–98; rainforest perceptions, 5–6; roads, 106–8; rubber extraction, 99–101; tree planting to increase rain, 21–22
Columbus, Christopher, 21, 99
commons, tragedy of the, 246–47
community capitalism, 248
community-managed forests: burial and memorial sites, 249; community capitalism, 248; deforestation and, 249–50; Democratic Republic of the Congo, 266; Guatemala's Maya Forest, 250–56; Kenya, 267–70; Mexico, 247; Nepal, 244–46; New England, 163; sacred groves, 248–49; vs. tragedy of the commons, 246–47; value of, 248, 270; Wapichan people in Guyana, 231; Xaxli'p people in BC, 241. *See also* agroforestry

Congo basin: colonial plunder, 100–101; displacement of Bayaka people, 238–39; flying river from, 36, 41, 42; forest disturbance vs. forest loss, 145–46; pre-colonial domestication, 93; roads, 108. *See also* Congo-Brazzaville; Democratic Republic of the Congo
Congo-Brazzaville (Republic of the Congo), 95, 238
conifers, 28, 46, 173
Connecticut, 162
conservation. *See* agroforestry; community-managed forests; reforestation; regrowth; tree planting
Conservation International, 252
Cook, James, 74
Cook-Patton, Susan, 206
coppicing, 214
Corry, Stephen, 239
Cortés, Hernán, 97
Costa Rica, 183–84, 185
Courtois, Valérie, 240
Covey, Kristofer, 51
Crouzeilles, Renato, 207
Crowther, Thomas, 192, 193–94
Culag Community Woodland Trust, 176
culture, 248
curare, 98
Curtis, Philip, 153, 154, 157, 202
Czech Republic, 169

Dan Saga (Niger), 210–11, 213, 214
Darién swamp forest, 106–7
Darling, Patrick, 92
Dasgupta, Partha, 34, 80
data: Canadian interpretation issues, 147–51; deforestation and, 147, 151–53
Dayak people, 117–20, 240
Decade on Ecosystem Restoration, 212, 270
deciduous trees, 46, 173
deforestation: albedo cooling from, 28–29; Canada, 147–51; carbon emissions from, 26; Chaco region, 136–37; chainsaws and, 104; climate change and, 49–50; from colonialism, 103–4, 164–65; from commercial agriculture, 137–38; community management and, 249–50; corporate promises, 141–43; Costa Rica, 183–84; data issues, 147, 151–53; economic development and, 186; in Europe post–cold war, 170; vs. forest disturbance, 145–46; Indonesia, 104, 113–14; New York Declaration on Forests and, 140–41, 142–43; rainfall decline from, 23–24, 37–39, 42; recovery of, 96; Sarawak, 115–20; savannization from, 62–64, 64–66, 76, 80; small-scale farmers blamed for, 144–46; temperature changes from, 25; US, 273. *See also* logging and timber market
"degraded" forests, 155–57, 205–6
DellaSala, Dominick, 76
Democratic Republic of the Congo, 189, 220, 239, 266. *See also* Congo basin
Denevan, William, 89
Denmark, 172
deserts: albedo, 28; reforestation and rainfall, 40; tipping points towards, 80; Yatir Forest in Israel, 195–200
Devine, Jennifer, 251, 254, 255–56
Dickman, Chris, 74
diplomacy, hydro-, 41–42
Diver, Sibyl, 241
Dogwood Alliance, 133
domestication, forest: Amazon, 85–88; Congo, 93; as global phenomenon, 7, 90; Maya, 90–91; North America, 89–90, 164; pristine forest myth and, 86, 88–89, 163;

Southeast Asia, 91; Sungbo's Eredo, 91–92; *terra preta,* 93–96
Dooley, Kate, 171–72
Drax power station (UK), 130–33, 134
Dubois, Clare, 187
Duta Palma Group, 113

economic development, 185–86
Ecosia, 187
Ecuador, 1–4, 34
Edwards, David, 156
electricity. *See* biomass-burning power plants
Ellis, Erle, 96, 193
El Niño, 41
El Salvador, 155, 185, 204–5
Emmerson, Archie Aldis "Red," 165
Eni, 194
Environmental Investigation Agency (EIA), 125, 126
Enviva, 133
Erickson, Clark, 87, 88
Essissima, Joseph, 217
Estonia, 249
Ethiopia, 41, 95, 146, 189, 220–21
Europe: acid rain, 168–69; agricultural decline and reforestation, 170–72; carbon and, 173; cold war, 169–70; deforestation, 103, 170; European Green Deal, 173; fires in Mediterranean, 76; illegal timber imports, 125–26; monoculture forestry, 172–73; reforestation and regrowth, 10, 172, 174; responsibility for global deforestation, 186; sacred groves, 249; urban trees, 275. *See also* biomass-burning power plants; *specific countries*

Fairhead, James, 144–45
Farm Cove Community Forest, 163
Farmer Managed Natural Regeneration (FMNR), 214, 220
farming. *See* agriculture; agroforestry
Fawcett, Percy, 59–60
Fay, Michael, 93
Fearnside, Philip, 106
Felix, Tessa, 227–28, 228–29, 230
Fern, 125–26
fires: Australia, 72–75, 79, 80–81; controlled burns vs. suppression, 77–80; global pattern, 75–76; synergistic relationship with forests, 72, 74, 76–77; tipping points towards savannization and desert, 63, 80
Firestone, Harvey, 102
Firestone plantation (Liberia), 102–3
Fishlake National Forest, 17
Fitri Pauzi, Mohamad, 119–20
flying rivers: Amazon, 30–34; deforestation and, 37–39, 42; Eurasia, 36, 52; as global phenomenon, 35–37; international diplomacy and, 41–42; reforestation for, 39, 40. *See also* biotic pump
Food and Agriculture Organization, 147
Ford, Henry, 102
Fordlândia, 102
forest bathing (*shinrin-yoku*), 274
forests: albedo, 27–29; approach to, 7–9, 11–13; author's childhood experience in, 4; cooling effects, 25–26, 172, 272–73; as cooperative systems, 19; culture and, 248; intact, 154; long-distance influences of, 40–41; monoculture plantations, 114, 132–33, 134, 172–73, 189–90, 220; perceptions of, 4–5, 6–7; rainfall from, 21–23, 24–25; resilience of, 11, 81–82, 96, 209; roads and, 106–8; synergistic relationship with fire, 72, 74, 76–77; tipping points, 66; value of, 9. *See also* biotic pump; carbon and

carbon emissions; colonialism; community-managed forests; deforestation; domestication, forest; flying rivers; Indigenous peoples; logging and timber market; reforestation; regrowth; rewilding; tree planting; trees
Forest Stewardship Council (FSC), 253
forest transition, 185. *See also* reforestation
fortress conservation, 238
Foster, David, 163
France, 172
Fraser, James, 94, 95
Fredericks, Nicholas, 228
fungi, 18–19

Gabon, 126, 146
Galvão, Ricardo, 68
gao tree, 213
Gardiner, Will, 134
Garrity, Dennis, 214
Gaschak, Sergey, 208–9
Gatti, Luciana, 65–66
Gautam, Homnath, 245
Gauthier, Sylvie, 151
Germany, 168–69, 174, 249
Ghana, 95, 125, 216–17, 237, 262–65, 266
Ghazoul, Jaboury, 190
Ghosh, Subimal, 38
giant sequoia trees, 78, 164, 249
Giron, Juan, 254
Gitau, Simon, 269
Global Canopy, 142
Global Forest Resources Assessment (FRA), 147, 151–52, 154
Global Forest Watch (GFW), 147, 149, 151, 153
global greening, 26, 27
Global Partnership on Forest and Landscape Restoration, 202–3
global warming. *See* climate change
Gockowski, Jim, 217
Gomes, Patrick, 222–23, 231
Goodyear, Charles, 99
Gorham (NH), 163
gorillas, mountain, 155
Gorinsky, Conrad, 232
Gornitz, Vivien, 103–4
Gorshkov, Victor, 52, 54
Grain for Green program, 179, 181
Grande Muraille Verte, la, 211–12
Grandia, Liza, 252
grasslands: albedo, 28; Crowther's forest restoration assessment and, 193; savannization, 62–64, 64–66, 76, 80; value of, 200
Great Green Wall, 181–82
Green Belt Movement, 188, 267
green burials, 249
Greene, Dale, 131, 132
greenheart tree, 125, 232
Green Revolution, 218
greenwashing, 141, 194–95
Griffiths, Tom, 230
Griscom, Bronson, 191, 265
Guatemala: Maya Forest, 250–56
Guevara, Patricia, 4
Guinea, 95, 144–45
Guyana, 125, 227–34

Haggith, Mandy, 176, 177–78
Hamis, Mohamed, 261
Hansen, Matthew, 153
Hardin, Garrett, 246
Hasan, Mohamad "Bob," 111
health benefits, from trees, 273–74
Hecht, Susanna, 204, 237
Heckenberger, Michael, 86–87
Helvetas, 245
Hernández, Francisco, 97–98
Herzog, Werner, 6
Ho, Ming, 127
Hodnett, Martin, 57
Holl, Karen, 166, 206
Holmgren, Peter, 216

Hong Kong, 272–73, 274
Hooker, Joseph, 22, 40, 101
Houghton, Richard, 203
Hua, Fangyuan, 181
Humboldt, Alexander von, 5–6
Hundertwasser House (Vienna), 275
Hunt, Chris, 91
hydroxyl, 48

Ijebu kingdom, 91–92
Ikombo, Benjamin, 215
Imberger, Jörg, 24
India: Bonn Challenge and, 189; flying rivers and, 38; Forest Rights Act, 250; reforestation, 10, 185, 186; sacred groves, 249; TreeSisters and, 187
Indigenous Leadership Initiative, 240
Indigenous peoples: Amazon, 234–37; biodiversity and environmental protection by, 88, 237, 239–40, 270; Canada, 240–42; control over forests, 248; displacement from protected areas, 238–39; fires and, 77–79; US, 242–43; Wapichan in Guyana, 227–34. *See also* domestication, forest
Indigenous Peoples Burning Network, 79
Indonesia: Bonn Challenge and, 189; deforestation, 104, 113–14; "degraded" forests, 156; fires, 75; highways, 107; illegal logging, 111–13, 127; moratorium on deforestation practices, 114–15; oil palm cultivation, 114; pre-colonial domestication, 91; pulp and paper industry, 113–14; rainfall decline, 39; Transmigration Program, 105. *See also* Borneo
Inhofe, Jim, 256
Innu, 240
Inslee, Jay, 80
intact forests, 154
Intergovernmental Panel on Climate Change (IPCC), 29, 191, 205–6
international diplomacy, 41–42
Iran, 183
Ishir people, 139
isoprenes, 46, 48. *See also* volatile organic compounds
Israel, 195–200
Italy, 171
Ivory Coast, 146, 216–17, 249, 265

Jagan, Cheddi, 230
Jaja (Tanzania), 261–62
Jambai Anak Jali, 119–20
James, Ron, 230
James, Tony, 230, 234
Japan, 104, 133, 246, 249
Jiang Gaoming, 182
Jiang, Hong, 183
Johnny, Angelbert, 230
Jones, Matthew, 75–76
Jost, Lou, 1–3, 4
Judaism, 5
Jung, Carl, 5

Kanuku Mountains (Guyana), 233
Karuk people, 77, 79
Karungari, Sarah, 268
Kauppi, Pekka, 185
Kayapo people, 235
Kayong, Bill, 117, 118–19, 120
Kazakhstan, 183
Kenya, 145, 187–88, 215–16, 249, 267–70
Keren Kayemeth LeIsrael–Jewish National Fund (KKL–JNF), 196, 198, 199–200
Kew Gardens, 101
Keys, Patrick, 37–38, 41
Kikuyu people, 249
Kingley Bottom (UK), 5, 276
Kiyonga, Dia, 261–62

Klinger, Lee, 77, 164
Klöckner, Julia, 173
Knight, Greg, 132
Kolden, Crystal, 79, 80
Königsbrücker Heath (Germany), 174
Kouwenhoven, Guus, 122
Kremen, Claire, 218
Kurganova, Irina, 171

La Condamine, Charles-Marie de, 98, 99
Laguna del Tigre National Park, 251
Lake, Frank, 79
Landry, Jean-Sébastien, 148, 150
larch, 273
La Rose, Claudine, 231
latex, 99, 100–103, 232, 253
Latin America, 97–100, 153, 248. *See also* Amazon; *specific countries*
Latvia, 171
Laurance, William (Bill), 105, 107, 108
Lawson, Sam, 123, 126
Leach, Gerald, 145
Leach, Melissa, 144–45
Lee, Kenneth, 87, 88
Lee, Stephen, 117, 119, 120
Legat, Stanislav, 129
Leopold, Aldo, 78
Leopold II (king of Belgium), 100–101
Lewis, Simon, 27, 189–90, 194, 220
Li, Qing, 274
Liberia, 95, 102–3, 121–23, 265
Linnaeus, Carl, 98
Lister, Paul, 177
Loess Plateau Watershed Rehabilitation Project, 179–81
logging and timber market: for biomass-burning power plants, 127–29, 130–32, 170; Borneo, 109–11; Cambodia, 123–24; China in international timber market, 124–26; crackdown on illegal logging, 126–27; Ghana's chainsaw millers, 262–65, 266; global scope of illegal logging, 123; illegal imports to Europe, 125–26; Indonesia, 111–13; Liberia's logs of war, 121–23; Sarawak, 115–16; small-scale in Africa, 265–66. *See also* deforestation
London, 271
Longfellow, Henry, 163
Lula da Silva, Luiz, 67
Lüneburg Heath (Germany), 174
Luo people, 215

Maathai, Wangari, 187–88, 267
Mabey, Richard, 5, 17
MacCleery, Douglas, 165, 166
Machakos (Kenya), 215
Macintosh, Charles, 99
Mackenzie River, 37
Madagascar, 187
Maggi, Blairo, 60, 70
Maindo, Alphonse, 265, 266
Maine, 162, 163
Makarieva, Anastassia, 52–54, 57–58
malaria, 98, 101, 232
Malawi, 95, 214, 220
Malaysia, 101–2, 104. *See also* Sarawak
Mali, 211
mangroves, Rufiji delta (Tanzania), 257–62
Mann, Charles, 90
Manser, Bruno, 117
Mansi people, 249
Marengo, José, 31, 34, 54–55
Marland, Gregg, 191
Martin, Claude, 155
Martínez, Cándido, 139
Martu people, 79
Massachusetts, 162
Mau Forest (Kenya), 268, 269
Mauritania, 211
Maya, 90–91, 163, 249
Maya Forest (Maya Biosphere Reserve), 250–56

Maya nut tree, 253–54
McAlpine, Clive, 39
McCormack, Michael, 73
McNicol, Iain, 146, 152
McSweeney, Kendra, 255
medicine, 98, 213, 232
Medvigy, David, 41
memorial and burial sites, 249
Menchú, Rigoberta, 237
Mendes, Chico, 67, 236
Mennonites, 140
Menominee people, 242–43
meranti tree, 109–10, 115, 125
merbau tree, 125, 127
Messok-Dja National Park, 239
methane, 48, 49, 50–52
Mexico, 189, 218, 247, 248
Mfisini (Tanzania), 257, 260–61
Michelin, 102
Milan: Bosco Verticale apartments, 275
Miller, Gifford, 81
Minang, Peter, 152
Minin, Leonid, 122
Mississippi, 130–31, 132, 166
Mitchard, Edward, 146
Moi, Daniel arap, 267
Mokany, Karel, 155–56
monoculture plantations, 114, 132–33, 134, 172–73, 189–90, 220
Montezuma, 97
Moomaw, William, 132
Moon, Sun Myung, 138–39
Mortimore, Michael, 215
Moss, Gerard, 30–34
Moutinho, Paulo, 67
Mozambique, 145
Mugo, Emilio, 268, 269
Mulonga, Julie, 260
Mursyid Muhammad Ali, 112–13
Myanmar, 125, 188

National AIDS Memorial Grove, 249
Native Americans, 77–79, 164. *See also* Indigenous peoples
Natural Resources Canada, 151
The Nature Conservancy, 79, 192, 252, 265
Naumo, Aggrey, 269
Nelson, John, 239
Nepal, 185, 187, 244–46, 248
Netherlands, 174
New England, 162–63
New Generation Plantations, 190
New Guinea, 107, 125
New Hampshire, 162
New York City, 273, 275
New York Declaration on Forests, 140–41, 142–43
Niger, 11, 40, 210–11, 212–14, 215, 216, 220
Nigeria, 91–92, 145, 189
Nile, 41, 42
Nimmo, Dale, 74
Nittler, John, 252
Nobre, Antonio, 30–31, 33, 55
Nobre, Carlos, 25, 64–65, 68, 69–70
noise pollution, 273
Nolte, Christoph, 236
North America: albedo effect from deforestation, 29; deforestation, 103, 153; fires, 75, 76; forest regrowth in, 10; pre-Columbian domestication, 89–90, 164. *See also* Canada; United States of America
North Carolina, 26
Nouabalé-Ndoki National Park, 238
Ntilicha, Mathew, 259–60, 261

Ogiek people, 268, 269
oil palm, 93, 104, 114, 116–17, 141–42, 147
Ojibwe people, 242
O'Keefe, Phil, 145
okoumé tree, 125
Oldekop, Johan, 245–46
Oostvaardersplassen (Netherlands), 174
orangutans, 107, 109–10, 155

orchids, 1–3
Oren, Ram, 169
Oslisly, Richard, 93
Ostrom, Elinor, 246–47
Ouch, Leng, 123
oxygen, 20

Pakistan, 38, 183
palm oil, 93, 104, 114, 116–17, 141–42, 147
Panama, 185
Pan-American Highway, 106–7
Pan-Borneo Highway, 107
Pangala, Sunitha, 50–51
Paraguay, 135–37, 138–40
Paris Agreement (2015), 191, 220
Parr, Kate, 200
Parsons, James, 89
Paul, Supantha, 38
Pecore, Marshall, 243
Pennington, Toby, 136
Pennsylvania, 161–62
Peru, 98, 99, 105, 236
Peters, Charles M., 222, 239–40
Pheap, Oknha Try, 123–24
Philippines, 91, 104
photosynthesis, 20, 26
Piple-Pokhara community forest (Nepal), 244–45
Planck, Max, 57
plantations, monoculture, 114, 132–33, 134, 172–73, 189–90, 220
planting. *See* tree planting
Pliny the Elder, 21
Plockhugget, 173
poison, 98, 232, 276
Poland, 172
Polman, Paul, 141
population density and growth, 185, 213, 216, 236
Porter-Bolland, Luciana, 249
Portugal, 171
Potapov, Peter, 153, 154–55
Povlsen, Anders, 177
Preisler, Yakir, 197–98
Priestley, Joseph, 99
*Psilocybe*, 97
PT Bayas Biofuels, 114
Puerto Rico, 203–4, 207–8
Putz, Francis, 156
Pyne, Stephen, 77, 78

Qie, Lan, 26–27
Quaranda, Scot, 133
quebracho tree, 138–39
Quechua people, 99
quinine, 98, 101

rain: acid rain, 168–69; climate models for tropical rainfall, 55; deforestation and, 23–24, 37–39, 42; from flying rivers, 30–34, 35–37; from forests, 21–23, 24–25; international diplomacy and, 41–42; received wisdom on, 23–24; reforestation for, 39–40; volatile organic compounds and, 46–47, 48. *See also* biotic pump
Rainforest Alliance, 253, 254
Rajão, Raoni, 68
Raleigh, Walter, 5
Ramage, Benjamin, 156
ramin tree, 113
reforestation: China, 179–82, 200; economic development and, 185–86; vs. natural regrowth, 10, 206–7; for rainfall, 39–40; West and North Africa, 211–12. *See also* agroforestry; regrowth; rewilding; tree planting
Regreening Africa, 220
regrowth: agricultural decline and, 146–47, 165–66, 170–72, 204–5; Amazon, 201–2; Atlantic Forest in Brazil, 205; biodiversity and, 155–56, 207–8; carbon capture and,

205–6, 213; Chernobyl, 208–9; complexity of, 153–57; Costa Rica, 183–84; data challenges, 152–53; El Salvador, 204–5; Farmer Managed Natural Regeneration, 214, 220; lack of appreciation for, 209; Niger, 11, 210–11, 212–14, 220; Pennsylvania and New England, 161–63; potential for, 202–3; Puerto Rico, 203–4, 207–8; vs. reforestation, 10, 206–7; succession and, 203; trend towards, 10–11; US, 166–67, 206. *See also* agroforestry; reforestation; rewilding
Reid, Leighton, 184
Reij, Chris, 212–13, 220
Reiter, Guy, 243
religions, 5
renewable energy. *See* biomass-burning power plants
Reo, Nicholas, 242
Republic of the Congo (Congo-Brazzaville), 238. *See also* Congo basin
rewilding: debate on, 174; Oostvaardersplassen, 174; process of, 11; Scottish Highlands, 175–78. *See also* agroforestry; regrowth
Rezende, Camila, 205
Rhode Island, 162
Rhodes, Cecil, 107
Ricciardi, Vincent, 219
Rieley, Jack, 109, 111
Rights and Resources Initiative (RRI), 248, 249–50
Rinaudo, Tony, 214, 220
Ritchie, Bill, 175
roads, 106–8
Robbins, Margo, 79
Roberts, Patrick, 90–91, 96
Robertson, Gordon, 176
Robson, James, 218
Rodas, Oscar, 136–37, 139
Romania, 170
Roosevelt, Anna, 86
roots, 18–19
rosewood, 123, 125
Rossby waves, 41
Rothschild, Alon, 199–200
Royal Dutch Shell, 194
rubber, 99–103. *See also* latex
Rudel, Thomas, 203–4, 219
Rufiji delta (Tanzania), 257–62
Ruma (Tanzania), 261
Russia: agricultural decline, 171; biomass-burning power plants and, 133; deforestation, 147, 153; flying rivers and, 42, 52; Forest Code, 57–58; intact forests in, 154; timber sales to China, 124–25. *See also* Siberia

Sabah, 156
Sabo, Peter, 128–29
Sack, Fred, 20
sacred groves, 248–49
Sahara, 24, 36, 40, 212
Sahel, 37, 40, 41, 210, 213. *See also* Niger
Salati, Enéas, 33, 106
Salelie, Yusuph, 257–58, 260–61
Salguero, David, 254
Salincourt, Kate de, 144
Salles, Ricardo, 68
São Paulo, 32–33, 39
Sarawak: conflict with Dayak people over forests, 117–20; logging, 115–16; oil palm cultivation, 116–17
satellite data, 152–53
savannization, 62–64, 64–66, 76, 80
Savenije, Hubert, 35–36
Schuster, Richard, 237
science, women in, 49
Scotland, 175–78

Scott, Cat, 50
Sebangau swamp forest (Borneo), 109–11
Seddon, Nathalie, 195
Sengwer people, 269
sequoia trees, giant, 78, 164, 249
sesquiterpenes, 46. *See also* volatile organic compounds
Seymour, Frances, 248
Shankman, David, 182
sharers vs. sparers, 218–19
Sheil, Douglas, 55, 95–96, 239
shifting cultivation, 96, 146, 153, 221–23
*shinrin-yoku* (forest bathing), 274
Shofet, Jay, 196, 199
Siberia: albedo and, 28; fires, 75; flying rivers and, 36, 37, 52; illegal logging, 124; refugees from Stalin in, 6–7; sacred groves, 249
Sierra Leone, 95, 103–4, 249
Silva, Marina, 33–34, 67, 68, 71, 236
Silvério, Divino, 61–63
Simard, Suzanne, 18–19, 151
Simons, Tony, 219
slash-and-burn farming. *See* shifting cultivation
Slovakia, 127–29
Smoky Mountains, 47
soil: *terra preta*, 93–96
South America. *See* Latin America
Southeast Asia, 91, 95–96, 104, 153, 216. *See also* Borneo; *specific countries*
South Korea, 133, 185
soy, 60, 67, 68, 142, 219
Spain, 171, 172, 246
sparers vs. sharers, 218–19
Spracklen, Dominick, 21, 23, 24, 50
Spruce, Richard, 2, 6, 101
Stanley, Henry Morton, 6
Steer, Andrew, 250
stomata, 20, 46, 272
Strong, Maurice, 184
Stroud, Stedson, 23
suburbia, 163, 167
succession, 203
Suharto, 111
Sumatra, 25, 107, 111–14
Sumner, Janet, 148
Sungai Bekelit (Sarawak), 117–19
Sungbo's Eredo, 91–92
Sun Xiufang, 125
supply chains, 140–42
Survival International, 239
Suzano, 205
Sweden, 172–73
Switzerland, 246–47

Taib Mahmud, Abdul, 115–16
Tal, Alon, 197
Tan, Minghong, 182
Tanguro farm (Brazil), 60–61, 63–64, 66, 70, 234
Tanzania, 220, 257–62
Tappan, Gray, 213, 216
Taylor, Charles, 121–22
teak, 125, 126
technology: for carbon capture and storage, 134; Indigenous use of modern technology, 229, 230, 234
teleconnections, 41
temperature: albedo and, 27–29, 198; cooling effects of forests, 25–26, 172, 272–73; urban heat island effect, 272–73. *See also* climate change
Tempus, Alexandra, 243
Teo, Michael, 117
terpenes, 46. *See also* volatile organic compounds
*terra preta*, 93–96
Thailand, 91, 104, 185
Thompson, Jonathan, 162
tigers, 107

timber market. *See* logging and timber market
Timmermans, Frans, 173
tipping points, 64–66, 80
Tomich, Thomas, 222
Töpfer, Klaus, 237
tourism, 233
tragedy of the commons, 246–47
Trans-African Highway, 107
Trans-Papua Highway, 107
transpiration, 20–21, 25, 34, 53, 172
Trans-Sumatra Highway, 107
Trase, 142
tree planting: Bonn Challenge, 188–89, 190, 191–92, 200, 220; carbon accumulation and, 190–95; global popularity of, 187–88; to increase rainfall, 21–23; monoculture plantations, 114, 132–33, 134, 172–73, 189–90, 220; vs. natural regrowth, 206–7; in the wrong ecosystem, 200; Yatir Forest in Israel, 195–200
trees: as biggest and longest-living organisms, 17; carbon accumulation, 18, 26–27, 190–95; in cities, 273–75; environment shaped by, 18; health benefits, 273–74; methane from, 50–52; religions and, 5; roots and fungal networks, 18–19; stomata, 20, 46, 272; transpiration, 20–21, 25, 34, 53, 172; ubiquity of, 17–18; volatile organic compounds from, 44, 46–47, 47–50. *See also* forests
TreeSisters, 187
Tripp, Bill, 79
Trump, Donald, 192
Turkey, 183
Turner, Robert, 276
Twain, Mark, 99

Uganda, 145, 155
Ukraine, 170, 208–9
Ulrich, Bernhard, 168
Unger, Nadine, 48–50
Unification Church, 138–39
Unilever, 141–42
United Kingdom: Drax biomass-burning power station, 130–33, 134; Kingley Bottom, 5, 276; London, 271; national arboretum and memorial ground, 249; urban tree coverage, 275; Wandsworth Common, 271–72
United Nations: China and climate change initiatives, 183; Decade on Ecosystem Restoration, 212, 270; Development Programme, 238–39; Environment Programme, 238; Food and Agriculture Organization, 147; Intergovernmental Panel on Climate Change, 29, 191, 205–6; Liberia and, 122; on natural regrowth, 203
United States of America: agricultural decline and reforestation, 165–66; Amazonian impacts on, 40–41; conservation, 165; deforestation and exploitation, 147, 164–65, 273; fires, 77–79; Guatemala's Maya Forest and, 256; Indigenous land stewardship, 242–43; National AIDS Memorial Grove, 249; reforestation and regrowth, 161–63, 166–67, 206
Upper Xingu basin (Brazil), 60–64, 86–87
urban heat island effect, 272–73. *See also* cities

Van der Ent, Ruud, 36, 37, 39
Van Gisbergen, Indra, 125–26
vanilla, 98, 101, 216
Vermont, 162
Vidal, John, 189
Vienna: Hundertwasser House, 275
Vietnam, 102, 104, 185, 189
Vilela, Thaís, 106

volatile organic compounds (VOCs), 44, 46–47, 47–50
Waldes, Nicola, 109–11
Waller, Donald, 242, 243
Wandsworth Common (UK), 271–72
Wang, Yunxia, 201
Wang-Erlandsson, Lan, 37, 41–42
Wapichan people, 227–34
Warufiji people, 257–62
Watts, George, 241
Weng, Wei, 39
West Africa. *See* Africa
West African monsoon, 24, 35–36, 40
Western Ghats, 38, 186
Wetlands International, 207, 260
White, Andy, 250
Wickham, Henry, 101
Widodo, Joko, 114–15
Wilber, Charles, 23
wildfires. *See* fires
Wildlands League, 148
Wildlife Conservation Society (WCS), 148, 238, 239
Wilkinson, David, 22
Williams, Christopher, 29
Wilson, Edward, 7
wind. *See* biotic pump; flying rivers
Wit, Marieke, 264, 265
WOLF, 128
Wolfensohn, James, 180
women, in science, 49
wood-wide web, 18–19
Worl, Sara, 78
World Agroforestry Centre, 220
World Business Council for Sustainable Development, 190
World Economic Forum, 192
World Land Trust, 139
World Resources Institute (WRI), 147, 237, 249–50
Worldview Impact, 188
Wrigley, Rebecca, 206
WWF, 107, 190, 222, 233, 238–39, 253

Xaxli'p people, 241
Xingu Indigenous reserve (Brazil), 60, 64, 234, 235

Yakir, Dan, 198–99
Yanosky, Alberto, 139–40
Yatir Forest (Israel), 195–200
yews, 275–76
Yosef, Gil, 40

Zapotec people, 247
Zomer, Robert, 216